Journalism in Change

STUDIES IN COMMUNICATION AND POLITICS

Edited by Bogusława Dobek-Ostrowska and Michał Głowacki

VOL. 3

Notes on the quality assurance and peer review of this publication

Prior to publication, the quality of the work published in this series is reviewed by an external referee appointed by the editorship.

Gunnar Nygren / Bogusława Dobek-Ostrowska (eds.)

Journalism in Change

Journalistic Culture in Poland, Russia and Sweden

PETER LANG
EDITION

Bibliographic Information published by the Deutsche Nationalbibliothek
The Deutsche Nationalbibliothek lists this publication in the Deutsche Nationalbibliografie; detailed bibliographic data is available in the internet at http://dnb.d-nb.de.

This publication was financially supported by Södertörn University in Sweden.

Library of Congress Cataloging-in-Publication Data
Journalism in change / Gunnar Nygren, Bogusława Dobek-Ostrowska (editors)
pages cm. - (Studies in communication and politics)
Includes bibliographical references and index.
ISBN 978-3-631-64921-3
1. Journalism–Technological innovations. 2. Journalism–Social aspects. 3. Online journalism. 4. Digital media. I. Nygren, Gunnar, 1954- editor. II. Dobek-Ostrowska, Bogusława, editor.
PN4784.T34.J67 2015
070.4'3–dc23
2014047055

ISSN 21971684
ISBN 978-3-631-64921-3 (Print)
E-ISBN 978-3-653-03989-4 (E-Book)
DOI 10.3726/ 978-3-653-03989-4

Peter Lang Edition is an Imprint of Peter Lang GmbH.

Peter Lang – Frankfurt am Main · Bern · Bruxelles · New York · Oxford · Warszawa · Wien

This publication has been peer reviewed.

www.peterlang.com

Contents

Abbreviations

M	mean
N	number
SD	standard deviation

Introduction: Journalism professionalization and journalistic culture as a matter of research

Gunnar Nygren and
Bogusława Dobek-Ostrowska

Journalism is a kind of social invention. It was born and developed with industrialism, enlightened then and strove for democracy. The link between these factors is well described by Schudson:

> "Journalism is the business or practice of producing and disseminating information about contemporary affairs of general public interest and importance (…), normally presented as true and sincere to a dispersed and anonymous audience so as to publicly include the audience in a discourse taken to be publicly important." (Schudson, 2003:11)

Journalism is for society and to serve the public sphere, but it is also a business to create the necessary economic conditions. Journalism is supposed to be autonomous from state and to be able to act freely. According to liberal theory it shall be the fourth estate of power (Burke, 1989). Independence from political and economic pressure constitutes the role of journalism in a democratic society.

Today, journalism is in crisis. It would suffice to mention the development of a network society (Castells, 1996), new notions of media (media-like services), changes of traditional one-way communication towards more interaction, and a system of "many-to-many" (Jakubowicz, 2009). Business models of the media industry are under pressure; American researchers have just noted: "there is no such thing as the news industry anymore!" (Anderson et al., 2013). The professional roles of journalists are questioned by social media and by users who have become the producers of media content (Lewis, 2012).

Convergence creates new kinds of interactive media systems, and has a profound impact on the functioning of traditional media firms (printed press, television, and radio). The media workplace is changing in the same directions as in other industries –workers have to be flexible, the demands on re-skilling and multiskilling increases and commercial pressure is much heavier (Deuze, 2007; Quandt and Singer, 2009). At the same time ideals and values are sluggish, old ways of thinking clash with new demands in daily work. Journalistic culture is perhaps stronger than many spokesmen of convergence assume (Fenton, 2010; Witschge and Nygren, 2009).

This development is most visible in the U.S. and the Western part of Europe. There are both differences and similarities between countries and media systems; yet newspapers still flourish in big economies like India and China and traditional TV is the main media format in most countries. Globalization has also created a convergence in journalistic orientations and practices in different parts of the world. Traditional Western ideals of objectivity and impartiality seem to dominate in many newsrooms, and there are many similarities in professional routines and editorial processes (Hallin and Mancini, 2004, 2012; Hanitszch, 2007; McQuail, 2013; Waisbord, 2013). But still there are also many differences among journalists in their ways of being professional, rather reflections of societal influence more than from media organizations and professional norms. Journalism is still very national in many ways, still connected to history and political traditions (Weaver and Willnat, 2012 a; 2012b).

Main concepts – definitions

In journalism studies, scholars hold different approaches to journalism as a profession. Zelizer (2004) defines five sets of perspectives in the studies of journalism – as a profession, as an institution in society, as text (content), as people, and as a set of practices. These perspectives are not mutually exclusive; the project "Journalism in Change" covers at least three of them:

- Journalism as a profession: a sociological perspective on journalism covering issues like autonomy, professional standards and values.
- Journalism as people: who is today's journalist, and what does this tell us about the position of journalism?
- Journalism as a set of practices: how is journalism produced today, and how changing processes influence thinking among journalists?

Journalism is this area of human activity, which has changed dynamically during the last few decades (see more in Chapter 1). What is journalism today? One of the most important factors is technology that provokes many implications for both society and journalism (McQuail, 2013: 13). Waisbord (2013) argues for a need to "reinvent professionalism." He further shows dilemmas and ambiguities, and defines "the professional logic of journalism." In this context, the model of the three traditions of journalism presented by Donsbach – subjective, public service and commercial (2010:41), seem very interesting and useful for our studies.

Professional (journalistic) culture is a key concept, which has been used in the project. In social and humanistic research culture is a "whole way of being," common ideals and practices in a group that separate it from other groups. Culture is socially constructed, and is carried by the people living in the culture as both values and ideals and as tacit knowledge hidden in daily routines. So culture is not only a question of ideology, it is also visible in practice – in journalism it also materializes in the working processes. In the words of Zelizer:

> "For recognizing journalism as a culture – a complex web of meanings, rituals, conventions and symbol systems – and seeing journalists (...) as its facilitators offers a way to think about the phenomenon by accounting for its changing, often contradictory dimensions." (2005:198)

The journalistic culture is an arena where different ideologies and practices can compete and live side by side. Journalistic culture has some common traits, but also big differences. It can be visible from the global level to national journalistic cultures, down to cultures in different media companies. In the comparative research project "Worlds of Journalism" (Hanitzsch, 2007) the notion "journalistic culture" has been deconstructed into three levels of analysis where culture is articulated:

- At the *cognitive level*, journalists shape the world, the interpretation of news, and news work in general.
- *Journalistic ideals* – beliefs and values about the role of journalists, the relation to external power and owners (political and economic), the relation to the audience and the role of journalism in a new media environment, professional ethics among journalists.
- *Journalistic practices* – the daily work and what a journalist is supposed to do (multiskilling, newsroom organization), autonomy and decision processes (the grade of power in the work), norms and routines in the work (tacit knowledge), what the journalists think about changes in their workplace.

Hanitzsch (2007) presents three dimensions of journalistic culture, such as: institutional roles, epistemologies and ethical ideologies, which have been useful when creating research tools for our analysis. The purpose of the project "Journalism in Change" is to identify common parts of transnational journalistic culture, general changes in journalism in different media systems, as well as differences between the three countries. It is also possible to relate the results to national differences in history and culture, to analyze the relationship between globalization and national differences.

Literature – earlier studies

Studies on journalism are very well developed in the Anglo-Saxon world, and in some Western European countries (Germany, the Netherlands, Belgium, Switzerland). In recent years studies in this respect have also been very popular in Scandinavian countries, and one can speak about "a golden decade" in Nordic communication research with the leading position of Sweden (Fernández-Quijada, 2014). Central and Eastern Europe is lagging behind in this respect.

Research in Western countries

Most research on how journalism changes in the era of new media development is conducted in the U.S. and Western Europe (Mitchelstein and Boczkowski, 2009; Quandt and Singer, 2009). The results are seldom related to differences in media systems and in journalistic cultures; it is often taken for granted that these results are valid in all kinds of media systems. There is a lack of empirical results in comparative research about changes in journalistic cultures. Most research on journalists is being conducted on a national level, as for example "The American Journalist. News People Around the World" (Weaver et al., 2007) and "The Swedish Journalists" (Asp, 2007). State-of-the-art includes many important chapters published in "The Routledge Companion to News and Journalism" (2010), where one can find an interesting consideration of professional identities (Donsbach, 2010) and journalism development (Waisbord, 2010). In this book, we note some arguments for a stronger homogeneity and convergence among journalists both nationally and internationally – due to commercialization and the increasing of editorial control (Donsbach, 2010).

During the last two years several books were published. The key concept of journalism, dilemmas and ambiguities of professional identity, logic of journalism, hybrid professional culture, post-professional journalism, were reconceptualized by Waisbord in 2013. One cannot forget about McQuail and his latest monograph "Journalism and Society" (2013), where technological changes in journalism are widely analyzed. It is worth mentioning "The Hybrid Media system" by Chadwick (2013), in which the author dedicates one chapter to changes in journalism due to technological conditions, and shows the boundaries between "professional journalism" and "amateur" blogging.

One of the few exceptions of comparative journalism study is "The Global Journalist in the 21st Century" (Weaver and Willnat, 2012) covering the changes in journalism in 21 countries around the world. Research led by Weaver and Willnat includes the examples of Poland and Russia analyzed by Stępińska et al.

(2012) and Pasti et al. (2012) accordingly. Another significant project, "World of journalisms", is led by Hanitzsch et al. (2010) and presents results from 18 countries (Russia included). Finally, "Media Accountability and Transparency in Europe (MediaAcT)" the European Union's project in 2010–2013 was conducted. "MediaAcT" helps us to understand accountability cultures in 14 different countries; it includes the example of Poland (Fengler et al., 2014).

Research in Central and Eastern Europe

Without a doubt journalism studies in Central and Eastern Europe have not developed to a similar extent. Their results often cover the example of one country; due to the fact that the studies are often published only in national languages, the access to them is very difficult for scholars from other part of the world.

Similarly to this, there are also some regional comparative studies concerning Central Europe but they are rather narrow and the results are not widespread. The majority of studies are dedicated to political communication and media systems but they lack research on journalistic culture. During the last decades only some scholars from Russia (Hanitzsch et al., 2010; Weaver and Willnat, 2012), Poland (Weaver and Willnat, 2012; Fengler et al., 2014), Bulgaria (Hanitzsch et al., 2010), Romania (Hanitzsch et al., 2010; Fengler et al., 2014), Hungary and Slovenia (Weaver and Willnat, 2012), Estonia (Fengler et al., 2014) participated in international comparative projects.

In recent years some new important publications have enriched studies dedicated to this region of Europe. One of them is "Comparing media systems beyond the Western world" edited by of Hallin and Mancini, where we find chapters on Poland (Dobek-Ostrowska, 2012), Lithuania (Balčytienė) and Russia (Vartanova, 2012).

"Media and Democracy in Central and Eastern Europe" (MDCEE), was an interdisciplinary project funded by the European Research Council (2009–2013). Many interesting reports and articles are the fruit of this research (Štětka, 2013; Bajomi-Lázár; Örnebring, 2013), but also the monograph of Bajomi-Lázár "Party Colonisation of the Media in Central and Eastern Europe" (2014). He analyzes five former communist countries (Hungary, Bulgaria, Poland, Romania, and Slovenia) and tries to explain variations in media freedom and the politicization of the news media in and across countries.

A collective book "Media Transformations in the Post-Communist World: Eastern Europe's Tortured Path to Change" (2013) edited by Gross and Jakubowicz, is another publication important for the "Journalism in Change" project, above all for the case of Poland. The authors present not only the positive effects of transformation after the collapse of communism, but also the problems and uncertainty of this process.

Finally, we should mention "Journalism that Matters. Views from Central and Eastern Europe" where we find one general study "How the Internet changes journalism: some trends in the 'West' and 'East'" of Bajomi-Lázár (2014), two chapters dedicated to Poland (Stępińska and Głowacki 2014; Milewski, Barczyszyn, and Lauk, 2014) and one to Russia (Pasti, 2014). All of them are very useful and enriched our research.

A transnational research project: "Journalism in Change: Professional journalistic culture in Poland, Russia, and Sweden"

The purpose of the project "Journalism in Change" is to identify common parts of a transnational journalistic culture and common changes in journalism in general in different media systems, as well as the differences among the three selected countries. It is also possible to relate the results to national differences in history and culture, to analyze the relationship between globalization and national differences.

The research design can be described as a "most-different" selection of cases. The project includes three countries representing different media systems, of different historical and political backgrounds and different sizes – Sweden, Russia and Poland, situated on the Baltic Sea. All of them have had relationships in the past. They were intense between Poland and Sweden in the times of the 16th and 18th centuries, and between Sweden and Russia from the 12th to 19th century, and Poland and Russia have had a very deep relationship from medieval times until today. The communist period (1945–1989) was significant for Polish and Russian journalism and professional cultures. Despite a common geographical location and history, the three states are different in many aspects: journalistic culture being influenced by different external factors, such as a democratic tradition (or lack of this experience), religion, education systems, economic development, and access to new technologies of communication. In fact, "Journalism in Change" is the first comparative project covering journalistic culture in these three countries.

We were aware of these differences from the beginning, but we also wanted to look at whether there are any similarities. With the study design it has become possible to analyze what changes in journalism in different types of society have in common, and what kind of differences come from the characteristics of each society.

Who takes part in the project and why?

Journalism has experienced deep changes in recent decades. For this reason, it seemed interesting to verify this general opinion in the case of only a few countries using empirical research. The points of departure for this book are based on two variables – technical and economic; it was our goal to observe how these two types of changes are influencing different media systems. The research project "Journalism in Change – professional journalistic cultures in Poland, Russia and Sweden" was conducted in the period 2011–2014. The project assumes a multidisciplinary approach, with researchers in journalism, media sociology, and political science. Researchers from Södertörn University (Sweden), Moscow State University (Russia) and University of Wrocław (Poland) worked together in the project to produce this final monograph. Two additional reports were published at the earlier stages of the project (Nygren et al., 2012; Anikina et al., 2013). Dissemination activities also include a number of articles published in scientific journals (Anikina, Dobek-Ostrowska and Nygren, 2013; Dobek-Ostrowska, Barczyszyn and Michel, 2013; Dobek-Ostrowska, Barczyszyn, Michel and Baranowski, 2013; Johansson, 2013, 2014; Johansson and Nygren, 2014; Nygren, 2012c).

Hypotheses

We formulated the two groups of hypotheses (see more 2.2.1.), which are important from the comparative perspective. The first group is linked with similarities/differences among journalistic cultures in three countries:

H1: *There is an increasing similarity in journalistic cultures in different media systems; market influence and liberal ideals are more common in pushing journalism in the direction of a commercialized Western model.*

H2: *The similarities between journalistic cultures are mostly superficial, nationally rooted traditions of history and culture still deeply influence journalistic cultures and preserve differences between them.*

The liberal ideals in journalism are getting stronger with market liberalism; a global media culture is emerging (Hallin and Mancini, 2004). This hypothesis can be defined as a homogenization of journalism in different media systems under the influence of technological and economic development, and the counterhypothesis can be the opposite that the similarities are mostly superficial:

H3: *Media development makes the profession weaker as a collective, the borders of journalism are more diffuse and professional autonomy weaker. Journalism, both as media content and as a profession, will be more difficult to perceive, because it will be different from that which we knew during the 20th century.*

H4: *Media development can strengthen the position of the individual journalist, giving him/her new possibilities both in research and in publishing. This can give journalists a new kind of autonomy.*

Other research shows that social institutions like journalism are hesitant to abandon their conventions even in the "age of the net" when communication patterns in society are changed (O'Sullivan and Heionen, 2008). A professional culture is sluggish, and moves only slowly in spite of changes in the surroundings – technical, economic and political. Journalists are often seen as conservative, and research shows that fast changes also promote a reaction of defense of old values (Witschge and Nygren, 2009). This can also be defined as hybridization, when hybrid systems emerge, melting together elements from the global development and national history and traditions (Hallin and Mancini, 2012).

Research questions

The hypotheses presented below provoke a long list of research questions, which are presented by the authors in each chapter. The research questions are linked with an area of analysis, but in general, three fundamental questions were addressed:

RQ1: What are the differences and similarities among journalists in Poland, Russia and Sweden when it comes to the basic dimensions (age, sex, education and professional training, membership of professional associations), working conditions, professional autonomy of the individual, organizational and societal level, ideals, standards and values of research's participants, their relationship with politics and politicians, their attitudes towards commercialization, the new technologies used in journalistic practice?

RQ2: What are the most important factors explaining the differences observed between media systems?

RQ3: How are the factors mentioned in RQ1 influenced by media development in the three countries?

Methods

The researchers from the three countries participated in the research workshops and in accomplishing the research. Three methods were used:

Survey/quantitative data analysis: A total 1500 respondents – a sample of 500 journalists from each country – Poland, Russia and Sweden, participated in the survey (see more 2.2.3.).

Interview/qualitative data analysis: 60 in-depth interviews were conducted with a broad selection of 20 journalists in each country (see more 2.2.4).

The survey and the interviews have covered several areas: 1) Who are the journalists? – age, gender and social position, income, and education. 2) The daily work – employment and conditions, perceived autonomy and influence. 3) Professional identity and relation to politics, commercialism and media owners. 4) Attitudes towards technology, interactivity and change in work. Social media use and multiskilling. 5) Professional roles in society, quality and press freedom.

Quantitative and qualitative comparative analysis: Surveys and interviews which were conducted in the three countries allowed us to use the received results for analyzing data sets by listing and counting all the combinations of variables observed in the data set. We compared the unique combination of values of its independent and dependent variables. We compared the data as numbers, percentages, standard deviation, means, factor analysis, and Pearson correlation.

The project has not studied journalism performance and media content. It has focused on the journalists, on how they think about their role in society and in media companies, about their daily work and their reflections on change. For example, journalists gave opinions on the quality of journalism, answering the question of whether it could be said to decrease, or not. There are no empirical data to support these opinions, no content analyses. The results are only the opinions of the journalists.

But in a comparative perspective, this still can bring new knowledge. It is possible to compare different generations, journalists in different kinds of media and in different media systems. What the journalists say has relevance, as long as we believe there is a connection between what you think and how you act.

Monograph "Journalism in Change"

This book is designed as a series of comparative chapters in different areas. Each author is responsible for the chapter, but the results have been discussed in the group and were carefully evaluated.

In Chapter One Gunnar Nygren gives a theoretical background to comparative journalism studies. The study covers theories on professions, autonomy, as well as research on how current media developments influence journalism.

In Chapter Two, background information on media systems in three selected countries is provided by Gunnar Nygren, Bogusława Dobek-Ostrowska, and Elena Johansson. The manuscript also contains a description of methods in the survey and the interviews and how the results have been analyzed.

In Chapter Three Michał Głowacki makes attempts to answer the question "Who is a journalist today?" He puts the emphasis on selected dimensions of comparative studies of journalism: demographic traits and facts on education, conditions of employment and the role professional associations.

In Chapter Four Jöran Hök analyzes daily work practices, working conditions, multiskilling and other dimensions of daily work.

In Chapter Five Gunnar Nygren focuses on the perceived autonomy among journalists and the degree of freedom within given frames in the three countries, as well as on political and commercial pressure on journalists in their daily work.

In Chapter Six Maria Anikina analyzes ideals and values, professional ethics and attitudes towards society. Also verification and other key values are analyzed in relation to media developments.

In Chapter Seven Bogusława Dobek-Ostrowska analyzes the relationship between journalism and politics, both the political preferences of journalists and how politics interferes in news processes.

In Chapter Eight Bogusława Dobek-Ostrowska analyzes the relationship between journalism and commercialization. This includes foreign ownership and also external economic pressure.

In Chapter Nine Elena Johansson analyzes how journalists relate to social media, how they use social media and for what purposes.

In Chapter Ten Gunnar Nygren and Bogusława Dobek-Ostrowska summarize the analysis, and relate the results to other comparative research in journalism. They discuss the questions of homogenization of journalism globally, or if development is more likely to be described as hybridization of journalism with new forms of media systems emerging.

The project has been led by Gunnar Nygren and financed by the Foundation for Baltic and East European Studies, linked with the academic infrastructure at Södertörns University in Sweden. The team of six scholars, supported by doctoral students and students from each of the studied countries, worked hard in order to achieve the outcomes and were intellectually fascinated by the job. The project helps us to know the situation of journalists and journalism, and to understand better how the countries developed journalistic cultures. The important question is how journalistic culture evolves today in a condition of intensive technological changes and what the perspectives are for the future.

1. Professionalization, media development, and comparative journalism studies

Gunnar Nygren

The concept of ongoing professionalization has been a key area in journalism studies for many years (Zelizer, 2004; Schudson, 2003; Waisbord, 2013). Journalism has been described as gaining influence within the media system, mainly in relation to political power and state. With common standards and professional institutions, journalism has grown stronger in modern society. But is this still the case in an emerging interactive network society? Is the development the same in different media systems – is it possible to still have a process of professionalization in some parts of the world, and the reverse process in other parts of the world; a de-professionalization? Is professionalization the same in different media systems?

This theoretical introduction gives some background to these questions covered in the project "Journalism in Change". It gives different perspectives from research in four areas:

- Professions and professional logic, based on sociological research on professions.
- Ideals and standards as described in comparative journalism research.
- Professional autonomy and pressure from the political and economic spheres.
- Professionalization and media development, professional control and open participation.

1.1 The professional logic

Firstly, we need to make a distinction between the two closely related concepts: "professionalism" and "professionalization". The word *professional* is often used to describe a skillful person, a person producing something of high quality. Professionalism is something that journalists embrace or pursue as a description of quality in their work (McQuail, 2010:286; Hanitzsch, 2009:416). These skills are something separating a professional from the amateur (the person who loves to do something, but still lacks the education and is doing so for fun).

The process of professionalization is something slightly different. Of course it has a connection to skills and to make a living from work. But in research, the notion describes when an occupation gradually develops into a profession that has all the characteristics of an independent profession. This process is described in sociology, where research has mostly analyzed classical professions, such as lawyers and medical doctors.

The sociologist Freidson has defined a profession as:

> "(…) A set of institutions, which permits the members of an occupation to make a living while controlling their own work." (Freidson 2001:17)

In his book *Professionalism, the third Logic: On the Practice of Knowledge* (Freidson, 2001:179) he defines three general kinds of logics to control work in areas demanding a qualified education:

- The bureaucratic logic – control by rules and laws. This control comes from state or organizations, and was analyzed by Max Weber on the basis of administrative structures in Germany.
- The marked logic – where control is based on market mechanism, where supply and demand decide the content and prize of labor. Consumers choose, and the producers adapt.
- The professional logic – where professionals control their work following standards and ethics formulated within the profession. This control is possible only if the profession has a knowledge monopoly, and can exclude those not fulfilling the demands of the profession.

In reality, these three kinds of ideal logics are always mixed, according to Freidson. But the models make it possible to discuss what kind of logic now prevails and how different ways to organize and control work lead to different kinds of practices. In the end, the struggle between the three logics is a question about power in daily work.

1.1.1 Division of labor

In general, professions developed as a result of the division of labor. Sociologists have defined two kinds of work specialization (Freidson, 2001:19–35). The first one is mechanical specialization: when the worker executes tasks decided by managers, a job with very little space for independent decisions. The second one – "discretionary" specialization: when the process demands that the worker makes independent evaluations and decisions, a job with the potential for innovation and creativity.

There is no sharp line between these two kinds of specialization; many kinds of work have both. However, the latter often demands formal training, and the skill of these workers is a mix of codified formal knowledge from school/

university and working knowledge from experience. This working knowledge can be both verbalized and a tacit knowledge that is not described in words. Professions developed from this discretionary specialization; autonomy in daily work was necessary for these kinds of professionals to be successful. This need for autonomy that is based in the working process has also been defined in journalism studies (Weaver et al., 2007:70).

Freidson defines the elements of the ideal type of classical profession giving this autonomy (2001:127–129):

- Specialized work, grounded in a body of knowledge and skill that is given special status in the labor force.
- A division of labor that is controlled by occupational negotiation.
- A sheltered position in the labor market based on the qualifying credentials created by the occupation.
- Training programs in higher education that are controlled by the occupation.
- An ideology that asserts greater commitment to doing good work than to financial gain, to the quality rather than the economic efficiency of work.

Together, these elements provide the profession with a kind of autonomy, both as a group in society and for individual members. Being part of a profession gives the individual professional power in his or her work. A profession as a group has influence and power in society in relation to other social groups.

This is the ideal model, but Freidson also admits that criticism against professionals have weakened the credibility of professionalism; it has made professions more vulnerable to market and bureaucratic forces and less resistant towards maximization of profit. Even more importantly – the professional ideology that claims the right and even the obligation to be independent has been undermined (Freidson, 2001:220).

1.1.2 A century of professionalization

Journalism is not a full profession, according to Freidson's criteria. During the 20th century, journalism developed into some kind of "semi-profession", according to journalism research (Weaver et al., 2007:71). Journalism has strengthened some of the elements of a profession – there is a specialized body of knowledge and formal training for journalists, there are professional standards and formal institutions of journalism in many countries, journalists struggle for independence and there is an ideology of journalism that is in service of society (Shoemaker and Reese, 1996:92–93).

In American journalism, professionalization was closely connected to commercialization. When the newspapers cut their connections to political parties, the occupation of "journalists" became organized through a set of self-governing professional norms and practices (Schudson, 2003:82–86). This group was also able to achieve some degree of autonomy from publishers.

In comparative studies of journalism, the degree of professionalization is one of the most important variables. In the analyses of media systems in 18 European and North American countries, Hallin and Mancini take professionalization as one of four variables. They focus on three dimensions of professionalization (Hallin and Mancini, 2004:34–37):

- The degree of autonomy and the control over the work process;
- Distinct profession norms such as news evaluation, ethical standards and professional integrity;
- Public service orientation – journalism as a public trust and not primarily as a source for profit for the owners

These dimensions connect closely to the classical theory on professions, described by Freidson. But still, they can be used mostly to analyze the direction of the development more than the actual status of "semi-professional" journalists. Most researchers conclude that journalism can't be a full profession that requires some kind of licensing – it would contravene freedom of speech to demand a license from those who want to express themselves in the media (Engblom, 2001:272, McQuail, 2010:288). There are also other researchers considering taking steps towards some kind of license for professional journalists as a way to strengthen professional autonomy. In Sweden, some kind of license was discussed when journalism education was established in the 1960´s (Furhoff, 1974:219). In their recent project on journalism in 14 European countries, the researchers conclude that this kind of accreditation might be a tool for balancing the competitive nature of the job market and the commercial interests of media organizations to use the growing number of poorly qualified workers in order to cut down on costs (Harro-Loit et al., 2012:153).

1.1.3 Two kinds of professionalism

There are different traditions in research on professions. In the 1960s the emphasis was (in the Anglo-American tradition) on analyzing the specific traits that constitute a profession, such as mechanisms to exclude outsiders from the profession. In the last 20 years, the emphasis has shifted to the dynamic processes through which occupations gain professional status. Now the important issues are not whether a profession fulfils all the conditions of an ideal typical profession, but about how professionalism is used to change and control an occupation. The process of professionalization is the most important: "the social process through which journalists struggle to claim professional status" (Schudson and Andersson: 2009:90). Research on this approach also considers the influences of specific cultural and historical traditions in different countries (Tumber and Prentoulis, 2005).

The British sociologists Evett and Aldridge (2003:399) explicate two different perspectives that are adopted when discussing professions: (1) as a normative value system created and upheld by the profession itself, giving the profession a collective identity – a positive interpretation; (2) as an ideology used as a mechanism for social order and discipline among professionals – a more negative interpretation.

The question is how the notion of professionalization is used in an occupation. Professionalization can come from within, when the professionals themselves can exert control and form the values and norms of the profession. It can also be used by outside groups (owners, state and organizations) to change an occupation and to use professionalization as a system of control and discipline. The latter interpretation can replace bureaucratic control, and can be used by external forces to promote change in the profession (Evett and Aldridge, 2003:404). In an analysis of the concept of professionalism in journalism, Evett and Aldridge (2003:561) conclude that the "professionalism" discourse is used as a tool to separate the producers from the product, and that change is legitimized through a reference to its allegedly "professional" nature. This can lead to arguments like "you have to do this job, because it is "professional". At the same time, the professionalism developing among journalists gives them their identity and belonging.

Journalism can be analyzed as an ongoing negotiation between these two sides of professionalism: between the organizational demands of standards, routines, and goals for the media company and the occupational professionalism – values, norms, and identity that are developing among journalists themselves (Örnebring, 2009). These two sides of professionalism also relate to autonomy for the profession, both on an individual level and for the profession as a group in media companies and in society. Autonomy for the individual journalists gets its strength from occupational professionalism, the values and identity rooted among colleagues. Autonomy for the profession as a group can be based on a mix of both kinds of professionalism – both as a defense against pressure from outside and as a way of maintaining professional order within the professional community.

This dual perspective on professionalization can be used to analyze change – what kind of changes are motivated by organizational arguments (economic, technical), and what kind of changes are motivated by occupational arguments (to defend and strengthen professional values). And how are the changes of this kind related to each other?

1.1.4 Professionalism in a global perspective

Research on professionalization in journalism has developed from a western tradition, very much based on empirical studies in the U.S. (Hanitzsch, 2009:416). Together with sociological research on professions, the perspective has been Anglo-Saxon and the criterions used to define professionalization have been set up by Western researchers. Different traditions have been defined, for example by Donsbach (2010:41): the subjective, the public service and the commercial traditions.

This makes it important to broaden the perspective. In the book *Reinventing professionalism*, Waibord analyzes professionalization from a global perspective (2013). His starting point is the classic research on professions, and he concludes that professional autonomy is challenged in general in late modernity. Journalism actively tries to redefine and strengthen the professional boundaries globally, and professionalism offers a way to protect news from governments and markets uninterested in critical rationality, accountability and transparency (Waibord, 2013:225).

It is also important to see the process an interaction between journalism and local/national traditions and institutions. Journalistic cultures have always been a mix of these, for example a blend of occupational news values and communitarian advocacy reporting (Waibord, 2013:231). These hybrid journalistic cultures develop in countries outside the western world, in Brazil, India and China. Still, Waibord concludes that professionalism is an important tool for autonomy of journalism in relation to state and markets:

> "Professionalism remains a critical and contested issue in contemporary journalism. It is not simply about normative horizons, occupational prerogatives or self-serving aspirations. Rather, professionalism refers to the constant redefinition of the occupational boundaries of journalism in a changing world." (2013:232)

1.2 Comparative journalism research

Journalism as an occupational field became professionalized at the same time it became commercialized as a business enterprise. It separated from political parties and grew more driven by commercial aspects and more organized by professional norms. This development came first in the U.S., but later in the UK and in the rest of Europe and in other parts of the world the process is still going on (Schudson, 2003:76). This might be the reason behind the fact that journalism studies developed as a separate academic field in the U.S. first of all with names as Walter Lippman and Robert Park in the 1920's (Zelizer, 2005:16–17). With this background, it is not strange that journalism studies often start from the U.S.model of journalism. In this model the notion "objectivity" grew among journalists as a professional standard (Schudson, 2001). From this, some traits typical for U.S. journalism emerged: event- action- and person-centred, negative, detached, technical and heavily dependent on official sources (Schudson, 2003:48).

Some professional traits have become similar for journalists in many parts of the world. This occupational ideology has been summarized into five traits by Deuze: public service – journalists work in the interest of the audience as watchdogs, active collectors and disseminators of information; objectivity – impartial, neutral, objective, fair, and credible; autonomy – free and independent in the work; immediacy – a sense of actuality and speed; ethics – a sense of ethics, validity, and legitimacy (Deuze, 2007:163).

1.2.1 Different roles for journalists

Even if the process of professionalization has been similar in many countries, this does not mean that professional norms and standards are the same. Comparative journalism studies have found both similarities and differences in journalistic values and roles. Different roles have been described as "active and participant" and "neutral and informative", as well as a kind of "interpretative" role between these two options (McQuail, 2010:283). These different roles are connected to national traditions and news cultures, for example American "objectivism" and a European continental tradition of commenting and a more literary kind of journalism. The researcher Chalaby calls journalism an "Anglo-American invention", when he describes the factual and information-centred kind of journalism that developed in the U.S. press, contrary to French journalism which is much less concerned with drawing a line between facts and commentary (Chalaby, 1996; Schudson, 2003:85).

Also other researchers have discussed the differences between the detached and non-partisan journalism in the U.S., and European journalism that is more involved in politics. German journalists have a greater desire to influence political processes, and a strong aspiration to increase their own participation

(Hanitzsch, 2009:417), in contrast to their American colleagues who are more neutral reporters. This can also be described as a difference between journalists who want to be active participants in social processes and journalists who prefer to be passive observers. Both these kinds of ideals can be regarded as different forms of journalistic professionalism. Also, the active participant journalist is seeking autonomy and working to serve society (Hallin and Mancini, 2004:41).

On a global level, Weaver found some common ideals during the 1990s. The role of getting the news out to people as fast as possible is common to most journalists, but other ideals – such as being a watchdog over those in power – were not as common in the 21 countries researched (Weaver and Wu, 1998). He found more disagreement than agreement on the roles of journalism in society, and the differences could be connected to political history and the current political situation. In the surveys there was some agreement among journalists in different countries on the importance of autonomy and on the protection of news sources, but in other areas there were big differences. As Weaver concludes, "no country or territory has a monopoly on professionalism among journalists" (Weaver and Wu, 1998:479).

1.2.2 Homogenization or hybridization?

Since the 1990s, the process of globalization has influenced journalism around the world. New global media, like satellite TV channels and media websites, have confronted audiences with new kinds of journalism, and the question is how this has influenced journalism. Some researchers describe it as a stronger influence on the Anglo-Saxon ideals of objectivity, separation of facts and comment and the ideal of the journalist as a distanced observer (Hallin and Mancini, 2004). This gets some support from a recent survey to 1 800 journalists in 18 countries to find "a general cultural understanding that is shared by most journalists around the world" (Hanitzsch et al., 2010:15).

However, there are still differences between a detached and neutral Western understanding of journalism and the more active promotion of values and social change connected to "development journalism" in developing countries and partly non-democratic countries. At the same time, the ideal of separation between facts and opinion among U.S. journalists might be changing as they put a stronger emphasis on interpretation in news coverage, for example on channels like Fox News. This means that old "objective" American journalism may also be changing.

Summarizing empirical findings from 33 countries, Weaver and Willnat cannot see any general convergence in global journalism in their last edition of "The

Global Journalist" (2012). There are many similarities concerning professional ideals, but there are also many differences reflecting societal differences (cultural and political) more than influences of media organizations and education. There is not any culture of global journalism yet, they conclude (2012:545).

In the book "Comparing Media Systems Beyond the Western World" Hallin and Mancini (2012) gather researchers from countries outside the 18 countries in Western Europe and North America giving the empirical base for their model. The researchers use the model of Hallin and Mancini to discuss the development in their own media systems, and in the final analysis Hallin and Mancini discuss the importance of abandoning the traditional Western-centrism of media studies. They use the term hybridization to describe that global developments, trends, and values are not only globalized but also adapted to national media systems. Development produces hybrid media systems where imported norms and values are transformed and adapted to local political structures and practices, not homogeneous media systems where everybody becomes "liberal" (Hallin and Mancini, 2012: 284–285).

The notion "hybrid media systems" is also used by Chadwick (2013). This new mix is not a blend of different national cultures, but a mixture of old and new media and different logics of old and new media. This mixture is visible in different forms around the world, and the common trait is not only a conflict and competition between old and new media logics but also interdependence between them. New media like social media and other forms of net-based media do not replace "legacy media", but the system is changing when the logics meet each other: transmission and reception of old media and circulation, recirculation and negotiation of new media (Chadwick, 2013:208).

1.2.3 What is possible to compare?

In a research conference in Moscow State University in November 2013, Paolo Mancini discussed what is possible to compare in journalism studies. He said they used categories from the West to analyze journalism outside the "liberal" model. One example was the notion of "accountability" that comes from the perspective of journalism as having responsibility for the common good – if this value is lacking, it is not possible to talk about accountability or the notion of "political parallelism" that builds on the assumption of having a stable political system with parties and national political power. This is often not the case outside the Western world (a keynote speech at the conference).

So what is possible to compare in journalism studies? Hanitzsch creates a model for comparative analyses of journalism (2007:371). From earlier research he constructed seven dimensions in journalistic culture covering both normative aspects and the function of journalism in society:

- Institutional roles (interventionism – whether a journalist should be an active participant in society, or merely a passive observer; power distance – whether journalism should be loyal, or critical towards power in society; market orientation – whether a journalist looks upon the audience as consumers, or as citizens);
- Epistemologies (objectivism – whether a journalist believes in telling the truth, or thinks that he/she alone can give subjective impressions; empiricism – whether a journalist emphasizes facts, or more analytical journalism);
- Ethical ideologies (relativism – whether a journalist thinks that ethical rules are important, or believes that every ethical decision is relative; and idealism – whether general ideals and morals are deemed to be important, or not).

These dimensions summarize much of the comparative research in journalism, and they can be a point of departure for comparisons even beyond the Western world. For the comparative analysis in the project "Journalism in Change", these dimensions have been a starting point in creating analytical tools to study journalism in the three countries in question (see Chapter 2.1.2).

1.3 Dimensions of professional autonomy

Professional autonomy for journalists is hard to define. It is a question of relationships between the profession as a group and power in society (external autonomy), but also as autonomy for journalists in their daily work in relation to the media organization (internal autonomy). Autonomy on these two levels is related, but not necessarily coordinated.

Bourdieu describes modernity as a process of differentiation into semiautonomous and increasingly specialized spheres of action (Benson and Neveu, 2005:3). He calls these spheres 'fields', and emphasizes that analyzing fields is to think relationally. Within each field develops a *habitus*, structures that organize practices and perceptions. This is subjective and structural, a kind of socialized subjectivity that keeps the field together. The agents in the field share a *doxa*, a universe of tacit presupposition that organizes action within the field.

For Bourdieu, journalism is a "very weak autonomous field" (2005:5). It has a position as a mediator between fields of politics, economy and other fields in society. These fields are closely intertwined, making journalism a field characterized by a high degree of heteronomy. But still Bourdieu underlines there are possibilities for journalism to gain some autonomy in relation to other fields by developing a specific cultural capital. With social capital (networks, institutions) and symbolic capital (values and prizes) developed within the profession, journalism can also achieve some autonomy for the field. Still, Bourdieu also sees journalism becoming more commercialized and close to the economic field (Bourdieu 2005:41–43). Market-driven journalism makes the field weaker in relation to economic fields.

Autonomy of the journalistic field has always been threatened both in relation to political and economic fields (Champange, 2005:49). Journalists have always had to produce under economic and political constraints. Economic and technological development has put the journalistic field under pressure, the cultural and symbolic capital developed within journalism is weak and the boundaries around the field are getting weaker, even if media companies still seek legitimacy in the symbolic capital of journalism.

Bourdieu's perspective on journalism puts the relationship between journalism and other fields into focus. The degree of autonomy for the field is crucial in analysing the role of journalism in society, and also the position of the individual journalist:

> "In short, the field opens up a new unit of analysis for media research: the entire universe of journalists and media organizations acting and reacting to one another". (Benson and Neveu: 2005:11)

1.3.1 External and internal autonomy

For journalism, autonomy was initially a question of external autonomy in relation to power in society. Development differed greatly due to historical circumstances. In countries with strong liberal traditions, newspapers could achieve an independent position earlier; for example in the U.S. and Sweden at the end of the 18th century. In other countries with a tradition of a strong state and a more authoritarian rule, this struggle for external autonomy has been a much longer process and in many cases is still going on. Important issues in the struggle for external autonomy have been censorship and direct control by state power.

Even if a degree of external autonomy is achieved, it is not given forever: pressure from external powers to control the media can increase in times of crisis. The means of this control may be laws and regulations, financial pressure and expanding PR machinery (Schudson, 2003:149; McQuail, 2010:324–325).

Internal autonomy for journalism is about the position of journalists in relation to the owners and other departments of the media company. Historically, the party press in Europe created strong links between media owners, parties and political power. The journalist was supposed to share the political beliefs of the newspaper (Hallin and Mancini, 2004:26–29). Internal autonomy can also be a question of financial pressure from owners and other departments in the company. "Market-driven journalism" can be a powerful constraint on internal autonomy for journalism when the "wall" between the newsroom and marketing department is broken (McManus, 2009:226–227).

The internal autonomy of journalism developed with the commercialization of news and the break-up of the party press. Together with a strong (at least in Western Europe) public service on TV and radio, the ideal of an independent journalist developed during the process of professionalization in the second half of the 20th century. In Sweden, this is manifested in the general agreement between the Swedish Union of Journalists and newspaper publishers, where the third paragraph states:

> "A member of staff may not be imposed upon to write against his/her conviction or to carry out humiliating assignments. When to decide what is 'humiliating', the Ethical Guidelines for press, radio and TV should be taken into consideration." (*General agreement*, 2010–2012)

This paragraph is very seldom applied in daily practice, but its very existence gives the individual journalist a certain degree of autonomy in relation to editors and the media company. It also emphasizes a strong linkage between internal autonomy, ethical standards, and responsibility.

Having common ethical standards and guidelines is one important trait of the profession. In Sweden, these have developed within the profession since 1878, to strengthen the credibility of and public trust in the media and journalism. The Press Ombudsman and the Press Council still have a strong position in "legacy media", providing common standards for professional ethics (von Krogh, 2008). But ethical standards are not enough, as journalists must have the power to follow these standards. Without a degree of internal autonomy, giving the individual journalist power in his or her work, the journalist cannot be held responsible for following ethical standards.

1.3.2 Three levels of autonomy

Professional autonomy can also be studies on different levels – as an ideal, as a perceived autonomy among professionals and as some kind of factual autonomy. In studies of journalism as a social field, some degree of autonomy is necessary to perceive journalism as a separate field (Benson and Neveu, 2005:4). Historical studies of journalism describe how the ideal of journalism developing as a fourth estate in this field was created. But these ideals of detachment and autonomy for journalism differ between media systems due to history and political/social development (Schudson and Anderson, 2009:93). The degree of perceived autonomy among journalists has been researched in surveys the last 40–50 years – both on a national level and in big comparative projects. This perceived autonomy focuses on decisions in daily work, relations to other groups in media companies and questions on the influence on media content (Weaver and Willnat, 2012:532). The factual degree of autonomy is more difficult to research; it is difficult to find reliable indicators of autonomy. Historical studies have researched how external autonomy for media have developed, and newsroom studies can give some important pictures on autonomy for journalists in daily work (Tuchman, 1983, Konow-Lund, 2013. But in the end, the degree of factual professional autonomy is something that varies over time and in different situations.

This provides a matrix of at least six different dimensions in studies of professional autonomy for journalism. The project "Journalism in Change" covers only some of them – it concentrates on perceived autonomy within the internal dimension of autonomy. This is researched with questions in the survey and interviews on how journalists perceive their degree of freedom in their daily work.

We also know from earlier studies that the degree of political parallelism is or has been high in Russia and Poland; the ties between media, journalism and politics are close and intimate (Vartanova, 2012:127, Dobek-Ostrowska, 2012a:33). Therefore, the analysis in the project also covers external autonomy as an ideal and as perceived by journalists in the three countries. These areas are covered by questions on the role of journalists in society and the degree of influence of external power on editorial decisions – the selection of subjects and decisions to publish (see Figure 1.1).

Figure 1.1: Six dimensions of autonomy and the areas covered in the project "Journalism in Change"

	External autonomy	Internal autonomy
Ideals	x	
Perceived autonomy	x	x
Factual autonomy		

Source: Author

1.3.3 Autonomy under pressure in daily work

It is difficult to measure professional autonomy. In surveys it is possible to research the opinions of journalists on their autonomy, both perceived autonomy in daily work and more generally about influence in the newsroom. By comparing changes in time or answers in different countries, it is possible to draw conclusions about differences in the perceived autonomy and evaluation of daily work among journalists.

Surveys among U.S. journalists that ask questions about autonomy in their daily work have been conducted four times 1971–2002 (Weaver et al., 2007). These surveys have covered all kinds of media and different areas of the U.S.; the last survey in 2002 conducted telephone interviews with 1149 journalists. Two questions related to perceived autonomy, and both these indicators show a clear decline since 1971. The share of reporters saying they have almost complete freedom to select stories has declined from 60 to 40 per cent, and the share that say they have almost complete freedom to decide the emphasis has declined from 76 to 42 per cent. The last survey was made 2013, and the degree of perceived autonomy continues to decline – now 33.6 per cent say they have "almost complete freedom" in selecting stories (Weaver and Willnat, 2014:11).

There are big differences between journalists – young journalists in big news organizations have much less influence over their work than experienced journalists in small newsrooms. Reporters in TV have less freedom than reporters in radio and on weeklies. Hallin and Mancini also noticed a decline in journalistic autonomy in the "liberal model", including the U.S., mostly associated with the increasing influence of business motives in media companies (Hallin and Mancini, 2004:277).

There are also differences between countries and media systems. These differences are clear in a survey comparing reporters in Russia and Sweden. In Russia, reporters more seldom choose their own ideas for reporting, and they also feel political pressure both from owners and politicians when choosing subjects, as well as political pressure as a common obstacle to publication. But still, both in Sweden and Russia, the professional interest of the journalist is the most important factor in selecting subjects in daily work, according to journalists (Nygren and Degtereva, 2012:7).

There can be many reasons for decreasing professional autonomy for journalists. Weaver mentions some trends that are linked to greater financial pressure within media companies: the commercialization of news and the erosion of the wall between business and journalism. In addition, new media technology and the increasing speed of news work have an influence:

> "(…) traditional professional values such as proportion, verification and relevance have given way to a perceived need to get an assertion into public circulation as quickly as possible." (Weaver et al., 2007:73)

Another kind of pressure on professional autonomy comes from media development, and the fact that journalists now have to compete with other kinds of information and content on the net – social media, blogs and websites offering alternative pictures of news events. Journalists have to handle many kinds of interactivity with their audience, regarding both new types of sources and the audience as an amplifier of media content when it is re-distributed through social networks such as Facebook and Twitter:

> "Journalists in a network must acknowledge that they will retain power only to the extent they share it; without facilitating the broad exchange, and not merely the delivery, of information, they will find themselves increasingly irrelevant to the conversation taking place around them" (Quandt and Singer, 2009:141)

Professional autonomy is under pressure from all the challenges in media development and the network society's creation of new patterns of communication. However, it is also possible that media development can create alternative public spheres and a new kind of journalism that can strengthen autonomy for parts of the profession, whilst other parts of the journalistic profession experience a diminishing autonomy.

1.3.4 Political and economic pressure

What kind of constraints are there on the professional autonomy of daily work? The survey of U.S. journalists from 2002 gives some answers (Weaver et al., 2007:76). The results indicated four groups of limitations for the individual journalist:

- From agents outside the news organization, for example sources.
- Professional conventions that most journalists follow, like ethical guidelines.
- Lack of resources and commercial demands in profit-making media companies.
- Editorial policies and processes in the news organization.

In the U.S. survey, the highest levels of perceived autonomy in daily work were found among reporters covering a beat (who have their own specialty) and reporters in small news organizations, mostly in weekly newspapers and in radio.

In Swedish surveys conducted in 1995 and 2005, journalists were asked if the influence on media content has changed for different groups over the last ten years. According to the answers, the influence of journalists had diminished significantly and the influence of advertisers, the audience and politicians had grown. Also internally, journalists feel themselves to be losing influence in favor of the advertising department and the management of the company. Still, the strongest influence comes from the editors – both when it concerns editorial policy and daily decisions (Asp and Johansson, 2007:166).

Both these surveys provide some clues about constraints in the autonomy of the individual journalist. There seems to be strong commercial pressure and limitations within the newsroom organization. There is also strong pressure from outside, from sources and the PR industry; this influence has grown rapidly (Dinan and Miller, 2009:250). The relationship between journalists and their sources is often described as a negotiation in which both parts have something to give and something to gain (Berkowitz, 2009:111). When the power balance between the two parts of this negotiation changes, it will influence the results. Stronger sources reduce the autonomy of a professional journalist when he/she becomes more dependent on the sources.

1.3.5 Political parallelism and autonomy

The relationship between journalism and politics has a long history. Many newspapers started as political voices for parties and movements, and newspapers in Europe are still often labelled with political connotations. Hallin and Mancini (2004:26–33) describe what they call political parallelism, which has been very strong in European media systems. This system has several components: the media content reflects distinct political orientations; there are organizational connections between the media and political movements, for example, in ownership; a tendency for journalists to be active in political life, alternative career paths shaped by their political affiliation; partisanship in the audiences who choose the media closest to their own opinions; journalistic practices with a more activist tradition and stronger emphasis on comments (Hallin and Mancini, 2004:28).

This kind of political parallelism has gradually become weaker in Europe, and at the same time journalism has become more professionalized. Ties between newspapers and parties are weaker in the western countries analyzed by Hallin and Mancini, and political influence on television has decreased with the growth of commercial television. However, this does not mean that political parallelism is dead – this dimension is still alive in different forms and levels.

In general, professionalization in journalism and political parallelism are in conflict with each other. Hallin and Mancini also see a kind of journalistic autonomy within a system of political parallelism; autonomy for the activist journalist within a system according to the tradition of European intellectuals (2004:41). So, there is no simple relationship between politics and journalism.

From an American perspective, Schudson comes to the conclusion that commercialization encouraged professionalization among journalists when news became a commodity and not a political tool. Professional standards and practices were developed by commercial newspapers, giving them independence in relation to parties and politics. But commercialism also offers constraints for journalism, causing a conflict with professional values (Schudson, 2003:133).

There is also a third possibility – a combination of strong political and financial pressure. The media system that is developing in China has a combination of strong political control in the media and strong commercial pressure in media companies. In this kind of system, commercialization gives no freedom for journalism, at least in the area of politics (Schudson, 2003:130). This is close to what Vartanova calls the "Eurasian media model" (2007:114). In this model, media companies have to combine different motivations, both commercial demands and non-market goals that are defined by political elites. It is difficult for the journalistic profession to get support for their autonomy from any commercial motivation, and the degree of political parallelism is high.

1.4 A profession under pressure?

Rapid media development influences the professional culture of journalists. There is a lot of research into how journalism is changing due to new media and new technology (Deuze, 2007; Michelstein and Boczkowski, 2009; Quandt and Singer, 2009; Lee-Wright, Philips and Witschge, 2012). A short overview based on this research shows trends that can both strengthen and weaken journalism as a profession:

Changes in journalistic practices:

Multiskilling is a word that describes journalistic work in contemporary newsrooms. Journalists must do a lot of the technical production and be able to handle the whole process, including photography, editing and writing/producing. This can give the individual journalist greater control of the work process, but also a stronger focus on production and less on research and verification (Lee-Wright and Philips, 2012:63–80).

The demands to produce are increasing, each journalist has to produce more and this limits research and working autonomy. Less of the work is outside the newsrooms; more is about recycling content already circulating in the content management systems (CMS) (Nygren, 2008b:92–93; Philips, 2012:86).

Internet and computer assisted reporting (CAR) provides journalists with access to many more sources than before. It is possible to aggregate and analyze information on a scale never seen before, and this can make journalists less dependent on traditional sources (Appelgren and Nygren, 2014).

The labor market for journalists is changing in the same way as the rest of the labor market – more temporary jobs, and outsourcing production to freelancers and production companies. This can make journalists more dependent on editors, but also gives strong journalists the ability to choose where to publish (Deuze, 2007:154–155; Nygren, 2008b:42–43).

Changing media markets:

In media companies, the wall between newsrooms and business has eroded, and journalists' work more in cooperation with other groups in the company. This can not only reduce the autonomy of journalism, but also increase the influence on journalists who cooperate with other departments in the media company (Gade, 2008, McManus, 2009:220).

Strong competition and commercial demands increase the need for effective production. This can make newsrooms more dependent on content produced by others, for example PR and other sources. The time for verification shrinks, especially in online journalism with constant deadlines (Philips, 2012:90).

New media companies and new markets can increase pluralism in the media sector. At the same time, concentration and consolidation among media companies is reducing pluralism. This is both good and bad for professional autonomy, which can benefit from many channels (Deuze, 2007: Nygren and Zuiderveld, 2011).

Unclear journalistic borders:

Public relations and communication is expanding, and many journalists partly work in this sector. The borders between PR and journalism are more blurred, limiting the role of journalists as watchdogs (Dinan and Miller, 2009).

Media content on the border between journalism and commercial messages is growing, with advertorials and product placement. For the audience it is often difficult to differentiate between journalism and other types of content. Notions as "content marketing" and "product placement" are becoming more important, changing the borders of journalism (Wahlund et al., 2013:171)

New kind of content is developing on the net and in social media; some could be regarded as journalism but without professional values and ethics. The border between producers and consumers is blurred – who is a journalist on the net? At the same time a new kind of autonomy is growing on the net, but it is not necessarily professional journalism (Singer et al, 2011).

There is no simple answer to how journalistic professional culture is influenced by media development. The trends mentioned above are some aspects of change – some of them can strengthen professional autonomy and ideals, in other aspects it can become weaker.

A summary of this could conclude that there are differences between the autonomy of the individual and of the profession as a group. Individual journalists might gain from many of the changes, but for the profession the borders become blurred with other professions and groups in media production. Deuze's book is titled *Media Work*, and journalism is only one part of this. He stresses the liquid character of work in media today, the fluid reality in daily work when professionals negotiate their identity in relation to a growing convergent media culture (2007:235).

Fundamental challenges

Under these trends, there are some fundamental changes challenging the journalistic profession. These can be described in three areas: professional control or open participation; journalism as products or as processes and services; restructuring of the industry.

Control over work is a key issue in professional logic. This control is challenged in a developing participatory culture in the media system – the roles of producers and consumers are increasingly mixing with the growth of user generated content (UGC), crowdsourcing and social media filled with UGC, and media links mixing all kind of material (Lewis, 2012:837). To keep the control, professionals are dealing with boundary work to maintain control, to consolidate professional ideology under new circumstances.

Part of this is the change from journalism being products for consumption on different occasions, to become processes and services with no start and no end (Karlsson, 2010). With no deadline, news services work 24 hours, 7 days a week to present the news flow for audiences and these audiences are to a greater extent than ever before taking part in the processes with UGC in different forms. Even if journalists control the processes and are legally and ethically responsible for what is published in news sites, these processes also expand outside media into social media platforms being both sources and channels for distribution for traditional media.

The restructuring of the industry is changing the environment where the professional journalistic role was developed. Thousands of jobs for journalists have disappeared in the U.S. and Europe in recent years, the business models are weak and the broad media are fragmentized and changed to niche media. Journalists have to be entrepreneurs to find ways to survive, flexible and temporary newsworkers competing to sell content to anyone (Deuze, 2007:233).

There has been a strong technological determinism among journalists, a feeling that new technology is the reason behind the changes. But research also emphasizes that technology is only a condition for change – the result depends on the social, cultural and institutional practices where new technology is embedded (Örnebring, 2010a:58). For example, multiskilling is possible thanks to technology, but how it is implemented depends on if it is used mainly to reduce costs or to give new creative opportunities (Deuze, 2007:142; Nygren, 2008:70). Often these aspects are difficult to separate, they are intertwined and show that the use of new technology is not determined in advance.

It is also important to remember the lesson from comparative journalism research, when global trends meet the national level. The discussion about the crisis for the journalistic profession originates often from the North American and Western European experiences, and the situation is different in other parts of the world. In Asia and Latin America, newspapers and TV flourish, and journalists are needed to fill all media outlets with content for a growing middle class. In many of these countries, journalists struggle for an increased professional autonomy to be able to serve the public interest (Waisbord, 2013:227).

1.5 Conclusion

In the title, the object of study is labelled as "professional journalistic cultures" in Poland, Russia and Sweden. We know from cultural studies that cultures are not fixed, but remain in a constant flux and develop under influences from outside – from other cultures and areas. In journalism this becomes clear in the study of the history of journalism; for example in Sweden journalism has developed since the 18th century under French, German, British and American influence (Gustafsson and Rydén, 2010). The question is not *if* but *how* journalistic cultures are changing; if globalization brings a more homogeneous journalistic culture, or if the development rather can be labelled as a hybridization where some element of global values and standards in journalism are adapted to national journalistic cultures (Hallin and Mancini, 2012:286).

"Culture" is one of the key notions in this project, and the other is "professional". To be a professional demands some kind of control over your own work, to have some kind of autonomy to follow the standards and values of the profession. Research in professions emphasize this autonomy as a question of power, about creating institutions making it possible for the members of the profession "to make a living while controlling their own work" (Freidson, 2001:17). Research has described a professionalization of journalism during the 20th century, but the question is how this is influenced by media development. Digital technology and commercialization are global trends, but how does this influence nationally rooted professional cultures? Are the consequences for professional autonomy the same in different kinds of media systems?

For example: in Western countries, some researchers describe a de-professionalization of journalism: more unclear borders around the profession, harder economic conditions for traditional media and professional control questioned by an interactive network society (Nygren, 2008b:168, Örnebring, 2010b:568, Waisbord, 2013:60). But at the same time, journalists and media in many developing countries struggle for more professional control in relation to authoritarian regimes and also use new kinds of media platforms to achieve this. In countries with authoritarian regimes, professionalization can still be a strategy towards a greater degree of press freedom (Harro-Loit et al., 2012:153).

In the project "Journalism in Change" we use theories on professionalization and comparative journalism research to analyze how professional cultures are influenced by media development. Our basic assumption is that the changes are not the same in countries as different as Poland, Russia and Sweden. But it is also likely to find similarities – and these might bring journalistic cultures closer to each other.

In the end there is also the underlying question about media content; how are standards and practices among journalists influencing journalistic content, which is so important for all citizens to construct their picture of the world? This project does not include analyses of media content, but the basic assumption is that journalists′ belief about standards and values and their professional practice also influence the results in newspapers, broadcasts on TV and radio and in online news sites.

That is why professional journalistic cultures are not only a question for journalists, but for society as a whole.

2. Professional journalistic cultures. Design and methods in the research

Gunnar Nygren,
Bogusława Dobek-Ostrowska and
Elena Johansson

The object of this study is professional cultures among journalists in three countries. The first question in the process is to decide who is a journalist – and the answer is not evident. Deuze (2007:141) describes contemporary journalism as only one kind of work in the broad media sector – an industry with unclear borders between different parts and media workers going in and out of different professional roles. He calls them "portfolio workers" going between assignment and professional roles, which become liquid. Other research describes journalism as a profession passing through on the way to other better paid jobs (Nygren, 2011:219; Pasti et al., 2012:280).

In this study, journalists are defined as people professionally producing content based on facts (not fiction or entertainment) for what we traditionally label as "mass media", in contemporary debate also called "legacy media": newspapers and magazines, TV and radio channels, online and digital formats connected to the traditional media industry. Also the growing number of "content producers" outside media companies is included; freelancers and those employed in production companies. This classic definition of a journalist has problems, for example the borders towards public relations, content marketing and towards entertainment in feature journalism. It also means that independent bloggers and people producing all kind of content produced for social media platforms are not included in the definition, even if this content has journalistic qualities and sometimes also can generate incomes making it professional in some sense.

It is likely that the definition of "journalists" is going to change, professional borders are not fixed forever and media development changes the industry quickly (Anderson et al., 2013). But still this old definition has relevance in relation to the content for daily media consumption. This old definition also makes it possible to connect to earlier research on journalists and their professional cultures in the coming analysis.

In this chapter we give the reasons behind the choice of Poland, Russia and Sweden, and some background on professional journalism in these three countries. We also describe the design of the project and the research questions to be answered. The methods in accomplishing the survey and interviews are described in detail.

2.1 Three very different countries

Behind the choice of Poland, Russia, and Sweden for the study are both scientific and financial reasons. We knew from the beginning that these three countries, even if they are geographically close to each other, are different in many ways: the size of the countries and cultures connected to their language varies enormously, from 9 million Swedes to 143 million Russians. History and culture has followed different paths, with Poland struggling as a nation both before and during the years of communism while Sweden has had independence for many hundreds of years. Both Russia and Poland have long experience of war and totalitarian regimes, while Sweden has had a more stable development of peace, democracy and political compromise. There will be more about nationally rooted backgrounds of journalism later in this chapter.

The choice of three very different countries is a result of a "most-different" strategy in the search for a common professional culture among journalists. With these different countries, any similarities between journalistic cultures are strong indicators on a common professional culture. The differences between the countries will probably make it easy to find differences also in journalism, both in ideals and practices. Almost 40 years of earlier comparative research in journalism has shown that news production is contingent on the cultural, political and historical contexts that shape the journalists work (Hanitzsch, 2009:413). From this, the "most-different" strategy can be useful in searching for similarities among journalists and analyzing these in relation to media development and globalization.

There are also clear limitations in the design of the study. With only three countries, it is not possible to draw general conclusions about differences between media systems. Compared with other studies like "Global journalist" covering 31 nations (Weaver and Willnat, 2012) and "World of Journalism" covering 18 nations (Hanitzsch, 2012), our study is limited. "Journalism in Change" is rather defined as three case studies of journalism in different kinds of media systems. On the other hand, our study in some regards goes deeper into the subject with a combination of methods – both a big representative survey and long interviews with a broad selection of journalists in the three countries. This makes it possible to analyze differences also within the systems – between types of media and between generations of journalists. This makes it also possible to relate differences in journalism between the countries to differences within the countries.

The financial reasons for the choice of countries are connected to funding. With a generous support from the Foundation for Baltic and East European Studies in Sweden it has been possible to accomplish the project in the years 2011–2014. The foundation supports research in social sciences and humanities related to Central and Eastern Europe, for example comparative studies of countries in the area.

2.1.1 Journalism in Poland, Russia and Sweden

In the first report from the project, earlier research on journalism in the three countries was mapped and a point of departure formulated (Nygren et al., 2012). Here we will give a short overview on media systems, journalism and the relation to politics in the three countries.

Poland, Russia, and Sweden belong to three different media models with different specifics of professional journalistic cultures. In the terms of Hallin and Mancini (2004:70) Sweden represents a classical Democratic Corporatist model with strong development of mass media, political parallelism, professionalism of journalists and state intervention, but with protection of press freedom. Sweden was in 1766 the first country to introduce a constitutional law where censorship was abolished and the freedom of the press guaranteed (Gustafsson and Rydén, 2010:41)). Sweden also has had a strong autonomous public service since the 1950's, which however has been challenged by the competition stemming from the effects of commercialization and new media platforms since the beginning of the 1990s (Nord and Grusell, 2012).

Sweden has traditionally been a newspaper country, with a strong presence of local and regional press. Even if reading has decreased, in 2012 Sweden was still at the top in Europe, with 69 per cent reading a newspaper on an average day (*Daily media reach in Sweden*, 2012). The newspaper market struggles with decreasing circulation, owner concentrations and smaller income from advertisements, resulting in shrinking newsrooms (Wadbring, 2012:45). Sweden was also came top when it comes to the Internet – 83 percent of all Swedes used the Internet in 2012 (*Media use in the EU*:10) – and TV-use with 83 percent watching TV on an average day, *Daily media reach in Sweden*, 2012).

Journalism has strong institutions in Sweden with a Union for a long time organizing about 90 percent of all journalists in the traditional media (see Chapter 3.5.). Sweden also has a strong system of ethical self-regulation in the traditional media. But there are also tendencies of a de-professionalization of journalism that could be traced to the unclear borders of the profession and higher commercial pressure (Nygren, 2008b).

The Polish media model is placed near the middle of the axis that separates the Polarized Pluralist and Liberal Media models, having a mixture of features of both patterns (Dobek-Ostrowska, 2012a:49). On the one hand it is specified by strong political parallelism of publicly owned radio and television, which have been controlled by politically affiliated bodies and serves as "the tool of political communication and propaganda of the ruling parties" (Dobek-Ostrowska, 2012a:41), above all in the 1990s and 2000s. On the other hand, the Polish print press sector has a high degree of foreign media ownership; this press is mainly profit oriented and politically neutral (Dobek, Barczyszyn, Michel and Baranowski, 2013:64).

Despite a significant state intervention in public broadcasting service, Poland has quite a high rank in the Press Freedom Index (see Figure 2.1). Such global trends as commercialization and tabloidization of media are obvious in Poland; it leads to a decline in the quality of journalism, which tends toward lower standards with sensational and a more entertainment style as a product (Dobek-Ostrowska, 2012a:49).

In media use, Poland is a typical TV nation, with 84 percent of the population watching TV every day (*Media use in the EU 2012*, 2013). The use of the Internet is lower, about 46 per cent every day and the daily reading of newspapers much lower – only 16 per cent of all Poles reads a newspaper daily. This puts the newspaper industry under hard pressure with a high level of foreign ownership especially in regional newspapers (Szynol, 2008:147).

Being "beyond the Western world" Russia does not fit into Hallin and Mancini's paradigm. Vartanova (2012:139) identifies the Russian media model as a Statist Commercial model or a Eurasian hybrid system where adopted elements from the Western market economy are combined along with non-market goals defined by political elites. In the traditional paternal character of the media – state relationship, the media still "play the role of an innocent and obedient child" (Vartanova, 2002:66; 2012:142).

Access to media sources varies significantly in Russia because this is quite a large country. It is better in bigger cities, while poorer people in the less technologically developed rural areas are usually able to watch only one or two of the national state channels: Channel One and Rossiya TV. They cannot afford (or do not want) newspaper subscriptions, because it is expensive and not an all-round cultural tradition now. According to surveys from 2010, 76 per cent of the population watch national channels regularly (White, 2011:340). These big national state controlled channels keep the big country together, and only 20 percent read national newspapers regularly and 24 per cent read local newspapers (White, 2011:340). The use of the Internet has expanded quickly in recent years, especially in cities like Moscow and St Petersburg, where 64–67 per cent of households use the Internet at least every week. In 2012 in the whole country 61 per cent used the Internet every week (*TNS Gallup Media*, 2014).

Modern Russian media in some way are under direct state supervision (state owned media) and indirectly also pressured from state-owned companies or through pressure on the owners, which may be based on personal relationships between government officials and private media owners (Kiriya and Degtereva, 2010). All those mentioned above are often hostile to press freedom. Russian professional journalistic organizations are weak. Journalists are gripped in a vice of both political and commercial pressure (Nygren and Degtereva, 2011).

2.1.2 Three clusters of professional culture

Professional journalistic cultures are a complex mix of journalistic values, practices, norms and media products. On the one hand, it tends to unify across the globe, but on the other hand varies on cultural diversities. At the same time journalists across the globe often share the same professional values, follow homonymous professional standards and use similar practices. Hanitzsch et al. (2010:15–16) identify three main clusters of journalistic culture: "Western journalism culture", "peripheral Western" and a third group of developing countries, transitional democracies, which tend to be rather non-democratic. Regarding to Hanitzsch's analysis, Sweden can be referred to the first group, Poland to the second and Russia to the last.

In Sweden, large quantitative surveys to all union connected journalists (Wiik, 2012:33) show that the ideals journalists relate to are mostly to be watchdogs and act as a "fourth estate", to explain in an understandable manner what is going on in society, serving as guardians of the citizens and to be objective in their reporting (Wiik, 2012:34). This rhymes well with Hanitzsch's definition of Western journalism culture (2010:15). At the same time, Swedish journalism also experiences a strong commercial influence. In a growing commercial market system with many diverse actors, media today are more dependent on advertisers than ever before, and are also competing with these actors for audience attention, in Sweden just like the rest of the western world (Wiik, 2012:35). Swedish journalists declare that the audience today has a much higher influence on the media than they had fifteen years ago (Asp, 2012:115).

In Western contexts, both aspects, acting as a watchdog of the government and a watchdog of business elites, tend to go hand in hand (Hanitzsch et al.,

2012) and the ideals explained earlier and this growing commercial market confirms this in the Swedish media system as well. To monitor society is to monitor all the strong forces, not only the political.

Polish journalism has a pro-Soviet past where the function of journalists consists mostly of propaganda, agitation and social mobilizing. However, in that time Polish journalism was the most liberal and open journalism in the Soviet bloc, both in official media and in a broad sector of non-official media (Dobek-Ostrowska, 2012a:30). It should be underlined, that Poles developed a strong tradition of oppositional movement and samizdat press during two centuries without having their own state and under forty-five years of Soviet control. These conditions resulted in Polish journalists being both highly politicized and professionalized (Curry, 1990:23–24). Jane Curry claims that:

> "Despite an official ideology that regarded the media as instruments of the party, Polish journalists developed a strong professional culture during the communist period." (1990: 23–24)

This situation can explain a presence of high political parallelism in Poland after the collapse of communism in 1989. Journalists have had a visible ideological preference as conservative, liberal, socialist, etc., but only few media served as a real instrument of political parties' and a propaganda tool. Sometimes some of them – such as a prestigious weekly magazine "Polityka" presented clear left ideals – criticize the leftist party and ridiculed their politicians.

Modern Polish journalists consider their profession as a service to the public. According to Jakubowicz (2002:210) they however provide partisan political viewpoints. Following the advocacy style, Polish media are in many cases opinionated and politically partial but without any stable alliances with political parties (Dobek-Ostrowska, 2012a:37). Polish journalism varies in its level, autonomy and skills. However, one can identify two groups of journalists at the top of the profession: an older generation of brilliant publicists and analysts, which represent "qualitative" journalism, and a younger generation of "paid amateurs" which is often less skilled and experienced (Dobek-Ostrowska, 2012a:39). One more feature of Polish journalism is its literary roots. "Publicist" writing has a long tradition in Poland, with reporters as Ryszard Kapuściński and many magazines giving space for literary reporting and essays.

Russia has a long historical tradition of journalism as a service to the state (Trakhtenberg, 2007:123). Russian journalism has been firmly linked with literature: the famous writers were at the same time well-known journalists/publicists (Zhirkov, 2001:23, 180, 185). In Soviet times journalists functioned on the one hand as propagandists and agitators (Pasti, 2005:92, 109) but on the other hand as "missioners" and "a moral leg of society" (Roudakova, 2009:415). Vartanova and Azhgikhina (2011:44, 70, 72) note that journalism in Russia is rather a mission, than a profession. Following literature-centric and humanistic traditions Russian and Soviet journalists helped ordinary people and even assisted some social reforms despite cruel censorship. This missionary function of Russian journalism has revived in recent times.

Pasti (2007:125, 126, 127) identifies three professional subcultures of contemporary Russian journalists: "Soviet generation", "transitional generation" and "new generation". The senior practitioners (who entered the profession in the Soviet era) are conservative, continuing nowadays to hold a view of journalism in which authorities take responsibility for supporting the social order. They strived to patronize their audience, perceiving it rather as the passive object of their social initiatives. The "transitional generation" is more orientated to the role of entertainers, perceiving the audience rather as media consumers. For them journalism is a type of public relations; they work for the interests of influential groups and persons in politics and business. Both generations reveal the traditional concept of journalism as a derivative of power and see their responsibility as being to the media owners and as observance of the rules of the game.

The last group is pragmatic, taking journalism as a temporal: to gain experiences, an influential connection and to get a well-paid position in the state service or business, mostly going to public relations services. They are tolerant of the practice of corrupt materials, avoidance of criticizing the authorities and attracted to journalism as show business, mostly TV and glamour press (Pasti, 2007:127). In her later research Pasti (2012:25) defined this phenomenon as "journalism being as a social lift".

Modern Russian journalism also differs by this literature-centric peculiarity and tendencies to personification. According to Hanitzsch et al. (2011:283, 282) Russian journalists have "the most favorable attitude towards providing analysis" and perceive themselves more as co-operators and supporters of government and official policy.

2.1.3 Journalism and politics

Overall, in the context of democratic standards we have three different models in the project "Journalism in Change". According to the *Democracy index* (2012:3–7), Sweden is an example of full democracy, Poland – flawed democracy and Russia – authoritarian regime[1] (see more 7.3.). Generally, the quality of democracy is reflected by media freedom. An analysis of the data Press Freedom Index from 2002 to 2013 helps us to conclude how democracy works in the context of three countries (see Figure 2.1).

Figure 2.1: Press Freedom Index 2002–2013 (rank)

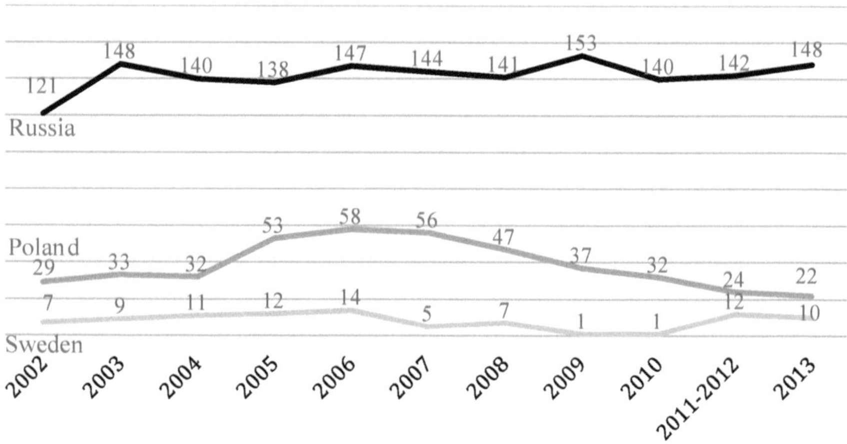

Source: http://en.rsf.org/spip.php?page=classement&id_rubrique=1054

The strong democratic values and standards in Sweden have created conditions for free media and journalistic professionalism. This country represents the Democratic Corporatist model and is being described, with other Nordic countries (such as Iceland, Finland, Norway and Denmark), as an ideal pattern of Democratic Corporatist tradition (Hallin and Mancini, 2004:70). The rank of Sweden in the Press Freedom Index from 2002–2013 (1st in 2009 and 2010, and the lowest 14th in 2006) confirms this high position. Political parallelism was very strong at the beginning of the 20th century when about 85 per cent of daily newspapers in Sweden had a clear political affiliation (Nord, 2008:103). On the other hand, the level of political parallelism diminished, since in the 2000s only 30 per cent of them were still linked with political parties and ideas. Nord underlines

1 According to *The Democracy Ranking of the Quality of Democracy 2013*, in 2011–2012 Sweden was ranked 2nd in the world (score 87), Poland – the 30th (score 71.1), Russia – the 95th rank (score 45.8).

a process of de-politicization and strong market orientation of Swedish media. He claims (2008:108) that state intervention, selective press subsidies and party connection obviously belong to the past.

Other research dedicated to the relationship between politicians and the media in Sweden shows that the role of journalists has grown systematically in political communication in recent decades. The surveys conducted by Strömbäck and Nord (2006:157) confirm that a perception of journalists' and politicians' power to influence politics and society is comparable and it is very high. The authors conclude that journalists rather than politicians exert control over news content. Additionally, what seems to be interesting in the context of our research, is that an interpretative style, which was rather absent in Swedish journalism for a few centuries, takes place more than a descriptive one.

All the data mentioned above show that the concept of the Democratic Corporatist model in the case of Sweden has changed and has rather historical significance. Sweden in the 2010s, with more social external influences, media convergence and digitalization, a growing role of the social networks (Bergström, 2011:256), drives at a more hybrid and liberal model of media.

In Poland, we historically found many features of typical Spanish or Italian models of media and politics at the beginning of social and political transformation after 1989. The high degree of ideological diversity and political conflicts was noted (Dobek-Ostrowska, 2012a:49). For two decades after the collapse of communism, many scholars perceived the level of the journalistic professionalism as lower than in the Democratic Corporatist and Liberal models. As far as autonomy, serving public interest and instrumentalization are concerned, it was much closer to the Polarized Pluralist model (Dobek-Ostrowska, 2012a:49; 2011a:25; 2011b:194–195). Despite the fact that some elements of professionalization had already appeared, the tradition of politicized journalism was still deeply rooted in Poland in the 2000s. The majority of journalists were far from being objective and represented biased political stances. Paradoxically, they were convinced that their civic responsibility demanded their personal involvement in the political course of events. Hadamik claims that it was a kind of combination of "the old and the new" elements of journalistic professionalism (2005:222). She believes that, on the one hand, a lot of values of journalistic culture still existed but, on the other hand, this culture evolved and modernized itself under the influence of global development trends.

Recently, the worst period for press freedom in Poland was between 2005–2007 (see Figure 2.1), when the conservative party of the Kaczyński brothers (Prawo i Sprawiedliwość/ Law and Justice) was in power. Poland has improved systematically its position after the parliamentary election in 2007 and presidential election in 2010. Nowadays, among post-communist European states it holds third position in the ranking (Estonia and the Czech Republic have occupied traditionally higher positions in the Press Freedom Index). After twenty-five years of democratic experience, it seems that not political pressure, but economic instrumentalization and commercialization are among the biggest risks for media quality (see Chapter 8). Not only in the case of Sweden, but also in the Polish case, a concept of a hybrid model (Dobek-Ostrowska, 2012a:49; Chadwick, 2013:22) and hybridization is very useful. Poland is in the group of new democracies, not ideal but a stable one, which has not lost all characteristics of the Polarized Pluralist model, but takes more and more features typical for the Liberal model, above all after 2010, and some for the Democratic Corporatist model (Dobek-Ostrowska, 2012a:49).

The Press Freedom Index confirms that an authoritarian model of media system in Russia still persists. We observe a significant difference in media freedom status between Russia and the other two states – Sweden (138 positions of difference) and Poland (126 positions of difference) – in 2012, when our research was conducted.

Vartanova (2012:129) says that during the first years after the collapse of the Soviet Union in 1991 (the first period of state-media relations), the Russian media accepted positively social changes as an alternative to the ideological monopoly and strong power of the communist party. She uses a concept of Roudakova (2008:43)"media-political clientelism" – a new phenomenon which emerged at that time. According to this framework not only the political elites were interested in the mass media, but also new players – emerging business elites. Both groups – political and economic, wanted to use the media as a good tool in achieving their aims. The second period, which started in the mid-1990s, was paradoxically a time of "the highest pluralism" with many oligarchic clans, financial and industrial groups, and regional powers. All of them wanted to instrumentalize the media. The third period started in the 2000s, and – as Vartanova claims (2012:133) – was initiated by Putin's "monocentric" political regime. She notes that:

> "(…) the media played a tacit role in the power system of Russia: While the state was exercising its policies to make TV less politically engaged, the advertising and media business easily filled "empty" niches of political programming with entertainment content." (Vartanova, 2012:134)

This gradual diminishing of political freedom of Russian media is noted by many scholars. It is a consequence of the increasing centralization of the Kremlin's authority and state regulation. According to Pasti, Czernysh, and Svitich (2012:267), the Russian media have drifted toward an "etatist" ideology. Political direct or indirect control over the media market has resulted negatively in the development of market and quality of journalism:

> "Although modern Russian journalism has transformed from being a state job (Soviet era) to a market freelance position (post-Soviet era), it is mostly unchanged in its political subordination. The paradox of market freedom and political non-freedom is a consequence of "guided democracy" or "simulation democracy" in which as Dmitry Furman (2010:11) explains, "democratic institutions and rules of law play a role of (fake) veneer, camouflage to hide the authoritarian system." (Pasti, Czernysh and Svitich, 2012:267)

This opinion has been confirmed by a survey presented by the Levada Centre in 2010, which indicates that more than 50 per cent of Russians shared an opinion that the news media in the contemporary country were a propaganda tool, and 66 per cent believed that they have been controlled by government authority or big business (Pasti, Czernysh and Svitich, 2012:268).

Bodrunova and Litvinenko (2012:40) use a concept labelled as a "politically hybrid system" to described the media model nowadays in Russia. This same term "hybrid" appeared in the case of Russia and Sweden but it is defined in another way. Russian scholars understand political hybridization as forms of convergent media practices, as a specific configuration of horizontal information and media agenda flows offline and online, redistribution of political influence between traditional and new media. In the case of authoritarian Russia, the Internet plays an important role in distribution of independent and plural opinions; in fact, it is one of the fundamental sources of political knowledge and values.

2.2 Design of the project

The project is briefly described in the Editors' introduction. Here follows a more detailed description of the design of the project, research questions, and some hypotheses.

The objects of study are professional journalistic cultures in Poland, Russia and Sweden. A culture has many levels and constructs a whole web of related values, practices and also things expressing this filled with meaning in relation to the culture (Zelizer, 2005:201; Hanitzsch: 2007:369). In this project we define culture on at least four levels:

(1) People – what kind of people are carrying this professional culture, sociological traits and demographic characteristics?
(2) Thinking: why are these people in the profession, what is important for them in their professional roles and what are the values and ideals they want to follow?
(3) Acting: how are these professional journalists working on different levels – on a basic level working conditions, tools and routines, and on the second one how they take place in relation to other professional groups in society and to their audiences?
(4) Symbols: how is this expressed in different ways, both in systems of symbols (prizes, prestige etc.) and in material ways like salaries and things connected to the profession?

All levels are bound together in a system we call professional culture. Of course this culture is not monolithic, there are also subcultures in different types of media and in different media companies. But we also assume there is a professional culture rooted in national history and traditions, following political and sociological traits of the country (Weaver and Willnat, 2012:544). This professional culture is not static, as no other culture either is defined forever. The national professional cultures of journalists are under a constant influence of globalization, media development and trends in journalism in other countries. But it is also under influence from internal changes in countries – political, economic and sociological. As Bourdieu (2005:3) emphasizes, social fields are always in relation to other fields, and this relation is a question of power, that is, of how strong economic, social, and cultural capital a field can gather. On the basis of this different kind of capital, social fields are in relation to each other and find their positions. The struggle to maintain borders and control the field is always going on – also for professional journalistic cultures.

To catch the traits of the professional journalistic cultures in the three countries, the following questions have been researched during the project:

1. Who are the journalists – age and gender, education and position in society, professional institutions like unions and ethical systems.
2. Why are they journalists – personal motivations, ideals and values on their role in society, beliefs behind their choice of journalism.
3. Practices – their daily work with traditions and routines, different kinds of tools, conditions in work and norms how to handle the gap between ideals and reality.
4. Autonomy – the degree of professional power over decisions in the daily work like choice of stories and emphasis in the stories. But also the role of political and economic influence on professional decisions – the relation to other social fields.

5. Influence of media development (new technology and commercial pressure):
   a) Changing practices and changing thinking about journalism?
   b) Is the position for the profession influenced – as a collective or/and as individual journalists? More or less power in the work?

With answers to these questions, it is possible to draw a picture of the professional journalistic cultures in these countries. These results are used in the final analysis.

2.2.1 Four hypotheses

Following previous research in journalism (see Chapter 1) two pairs of hypothesis are going to be tested in the analysis. They are partly contradictory, but not completely. The results can be contradictory and point in different directions. The first two concern the comparative perspective:

H1: *There is an increasing similarity in journalistic cultures in different media systems, market influence, and liberal ideals are more common in pushing journalism in the direction of a commercialized Western model.*

H2: *The similarities between journalistic cultures are mostly superficial, nationally rooted traditions of history and culture still influence journalistic cultures deeply and preserve differences between them.*

The answers to the first two hypotheses concern the relationship between similarities and differences, and this is a key question in the project; how is this relationship analyzed, in what direction is the development heading and at what speed?

The two other hypotheses concern media development and professionalization. Media development influences the professional cultures in all three countries, but the differences between media systems makes it likely that this influence has different results and different effects in the professional cultures. Two hypotheses can be posed:

H3: *Media development makes the profession weaker as a collective, the borders of journalism are more diffuse and professional autonomy weaker. Journalism as media content and as a profession will be more difficult to perceive as we know it from the 20th century.*

H4: *Media development can strengthen the position of the individual journalist, giving him/her new possibilities both in research and in publishing. This can give journalists a new kind of autonomy.*

The answers to the last two hypotheses concern the relationship between the process of professionalization and the opposite – a de-professionalization with an eroding professional culture and diminishing autonomy in relation to other social fields. The question is whether media development makes the journalistic profession stronger, or weaker, and whether there are any differences between media systems in this respect. These two hypotheses concern different levels of the profession – the collective and the individual – and the answers to them are not mutually dependent.

2.2.2 Beyond Western bias

There are some general methodological problems in comparative research, some of them summarized by Hanitzsch (2009:422–423). One major challenge is when researchers are doing some kind of "safari research", comparing other nations to their own by evaluating other cultures through their own value system. Another challenge is that variances within cultures can be greater than those across cultural borders.

These problems can be seen as a "Western bias" in research. To translate questions can be a way to get results possible to compare, but at the same time the meaning of the questions might differ between cultures. What is completely clear in Western journalism can be more unclear in other professional cultures. For example the notion "objectivity" can have many different meanings when it is transferred to other professional cultures. When norms are transferred from the Western Anglo-Saxon model, the meaning of them can be transformed and adapted to other circumstances (Hallin and Mancini, 2012:283–4). The result is then not a homogenization, but rather a kind of hybridization:

> "(….) the concept of journalistic professionalism needs to be unpacked and relativized – it has different dimensions that may not all develop in the same way. Professionalism also has taken different forms in different contexts; for example it has not always been associated with the "objectivity norm" that is central to its American variant." (Hallin and Mancini, 2012:290)

In Western bias, there is also a tacit understanding that the rest of the world should try to catch up with the norms developed in the western world. When it comes to Central and Eastern Europe this perspective has often been expressed in the notion of "transition" as a process when the post-communist countries develop to modern Western democracies. A quarter of a century after the victory of Solidarity in Poland and the falling of the Berlin Wall, it is clear that the process was not as simple as a short period of "transition" to market economy and democracy (Björklund and Rodin, 2009:11). Instead the results show a great variance, and national history and political structures play a great role in the development. This is clear for example in Russia, developing their own kind of "sovereign democracy" and political leaders emphasizing national traditions in a system with a strong state and authoritarian rule (White, 2011:355).

The project tries to handle these methodological problems in different ways. One key decision was to create a multinational group of researchers from three universities to work together during the whole project. The researchers from Södertörn University, University of Wrocław and Moscow State University met already during the planning of the project, and the different national perspectives have been present during the whole project. Another important feature of the group is the multidisciplinary approach – in the group there are researchers from journalism studies, political science and sociology of journalism.

To get under the surface of language and culture a combination of methods have been used during the project. Usually, comparative research is based on surveys (Weaver and Willnat: 2012; Hanitzsch, 2010). In "Journalism in Change" also long interviews with a broad selection of journalists in each country is used to get deeper in the questions. This combination of methods gives a deeper understanding of the results (for details about the methods, see below).

The project can be summarized in the following steps:

- In the first step, the researchers summarized research on journalistic cultures in the three countries to get a common ground for further work. This report was published in 2012 (Nygren et al., 2012).
- In the big survey the questions were discussed carefully among the researchers to avoid misunderstanding and problems in interpretation of the results. Some questions from other surveys were used to make it possible to compare results, but also other questions were given in the same areas to check the results.
- A broad selection was made of 20 journalists in each country for interviews each lasting one hour. These interviews were transcribed and translated into English to give all researchers access to them.
- The results from the survey and the interviews were discussed in several workshops during the end of the project. In these discussions interpretations were discussed and evaluated, and problems could be solved. The results from the surveys were published in a report in 2013 country by country (Anikina, Dobek-Ostrowska and Nygren, 2013).
- The results were also presented at several international conferences: "Future of Journalism" in 2013 in Cardiff, International Communication Association in 2014 in Seattle and Central and East European Communication and Media Conferences (CEECOM) in 2013 in Kaunas and 2014 in Wroclaw. In these presentations, the results were commented on by other researchers from many countries.

2.2.3 A survey of 1,500 journalists

The questionnaire contains 47 questions and was constructed on the basis of the six research areas (see Appendix 1). The questions were developed in cooperation between the researchers to reach the highest possible validity, so each question should be interpreted in the same way. Some questions have been used in earlier research (Weaver, 2007; Nygren, 2008; Nygren and Degtereva, 2012). In many cases different questions try to catch the opinions on one area to avoid problems with the interpretation, a kind of triangulation in putting the questions.

The survey was conducted during the spring and summer of 2012. The field-work was organized and carried out by teams in each of the three countries. There are no registers on journalists in the three countries, except in Sweden where earlier research on journalists has been based on the list of members in the Union of Journalists (Asp, 2012; Strömbäck et al., 2012). In the other two countries, the unions organize just a small part of the population and it was not possible to use the unions to get a representative sample. To use the same methods in the three countries, the project instead used quota sampling to select participants in the survey, to build a sample as representative as possible for journalists in each country (Lavrakas, 2013).

Drawing on previous knowledge about the media structure and where journalists work, a quota of journalists was decided for each media type with the goal of collecting 500 surveys in each country. Surveys were sent to a wide range of newsrooms in different parts of the country and different types of media, both on paper (mostly Sweden and Poland) and by e-mail with links to a web-based survey (mostly Poland and Russia). The survey was not sent to named journalists, but with the instructions to distribute the questionnaire to all journalists in the newsroom. The journalists returned the questionnaires themselves or filled in a web survey anonymously. When the quota for each media type was filled, the survey ended.

Details for each country

Poland – the number of journalists was estimated on the basis of "Almanac of Media" (*Almanach Mediów*, 2012) and phone calls to the editorial offices. This showed how many journalists that were needed from each group, and surveys were sent out to the editorial offices both as printed surveys and as mails with links to a web survey. Surveys were sent out until the required number was reached. In total more than 1000 printed surveys were sent out and a few hundred mails with links to the web survey.

Table 2.1: Basis for selection of Polish respondents

Media type	Percentage of journalists	Number of surveys
National newspaper	10	50
Regional newspaper	16	80
Magazine	24	120
Radio – channel (public service)	11	55
Radio – channel (commercial)	8	40
TV- channel (public service)	8	40
TV- channel (commercial)	9	45
Online media	3	15
News agency	3	10
Freelancer	8	45
All	100	500

Source: *In what kind of media company are you working today? (If you work for different kinds of companies, please give the main option).* The "Journalism in Change" Project Survey, 2012.

The sample is close to other surveys, for example in "Global Journalist" (Stępińska et al. 2012:260). The share of male journalists is 58 per cent in our survey, and 57 per cent in "Global Journalist". The median age is a bit older in "Journalism in Change", 37 years compared with 34 years in "Global Journalist". The share with an academic education in journalism is 38 per cent, compared with 32 per cent in "Global Journalist".

Russia – in Russia there are about 150,000 journalists, according to estimates from the mid-2000s. The main criterion for sampling was medium type. Official statistical information was used as a base for sampling. In 2011 there were fixed (registered and re-registered) more than 90,000 media: 67,727 printed media, 21,234 audiovisual media and 1,564 news agencies (Anikina, 2013:32) Online media formed the rest of the sample. From a geographical point of view the sample includes journalists from six federal districts (out of eight that Russia has) excluding the Siberian and North Caucasian federal districts. Respondents live in Central Russia, Southern, North-Western, Volga (Privolzhskiy), Ural and Far Eastern regions.

Table 2.2: Basis for selection of Russian respondents

Media type	Percentage of journalists	Number of respondents
Print media	67	335
TV and radio	21	105
Online media	8	40
News agencies	2	10
Production companies	2	10
Total:	100	500

Source: *In what kind of media company are you working today? (If you work for different kinds of companies, please give the main option).* The "Journalism in Change" Project Survey, 2012.

The sample is close to "Global Journalist" (Pasti et al. 2012:271–272) which used a combination of quota sample and members in the Union of Journalists: The share of male journalists is 38 per cent in our survey, and 40 per cent in "Global Journalist". There is a clear difference when it comes to age; in "Journalism in Change" the average age is 32 years, compared with 41 years in "Global Journalist". This difference is probably because one third of the population in "Global Journalist" consists of members in the Union of Journalists where the average age is quite high (in our survey over 40 years of age).

Sweden – earlier research on journalists at Gothenburg University (Asp et al., 2007) was used to define the proportion between respondents in different kinds of media. Newsrooms in all different categories in different parts of the country were chosen and the survey was distributed in paper together with an attached envelope to send in the answers. Surveys were also distributed to freelancers through cooperative groups of freelancers and also with the help of the Union of Journalists. In total about 1,400 surveys were distributed until the required number of respondents in each category was reached.

Table 2.3: Basis for selection of Swedish respondents

Media type	Percentage of journalists*	Percentage of respondents
Local/regional newspapers	32	29
Big city newspapers	14	14
Free newspapers	3	6
Magazines	17	13
SR (public service radio)	12	9
SVT (public service TV)	10	7
Commercial TV and radio	4	5
Production companies	1	2
News agencies & online only	6	6
Other	2	10**
Total:	100	100

* *Journalists employed by media companies in 2005. In this is not included about 13 % of all journalists working as freelancers 2005.*
** *Many of these are freelancers working mostly for magazines.*
Source: *In what kind of media company are you working today? (If you work for different kinds of companies, please give the main option).* The "Journalism in Change" Project Survey, 2012.

The sample is close to "Global Journalist" (Strömbäck et al., 2012:306). The share of men in our survey is 47 per cent compared with 50 per cent in "Global Journalist". In our survey, 38 per cent of journalists are older than 50, in "Global Journalist" the figure is 36 per cent. The answers in "Journalism in Change" comes from the whole country: 51 per cent from journalists in Stockholm, 10 per cent from the two other big cities Gothenburg and Malmo and 39 per cent from regional and local media. There are marginally more respondents from the three big cities than in earlier research – 61 per cent compared with 56 per cent in the latest survey to members in the Union (Asp, 2012:15).

The total sample of the survey "Journalism in Change" is described in Table 2. 4.

Table 2.4: The sample in the survey "Journalism in Change" 2012 (in per cent)

	Poland	Russia	Sweden	Total
Male	58	38	47	48
Female	42	62	53	52
Under 35 years	42	69	27	47
36–50 years	39	23	36	33
Over 51 years	18	7	38	21
Printed paid newspapers	26	35	42	34
Magazines	25	29	13	22
Public service/state radio and TV	22	10	16	16
	Poland	Russia	Sweden	Total
Commercial-free media*	25	21	12	19
Subcontractors**	3	6	17	8
Total number of surveys	500	500	500	1 500

* commercial TV and radio, free newspapers and online only
** news agencies, production companies and freelancers
Source: *In what kind of media company are you working today? (If you work for different kinds of companies, please give the main option).* The "Journalism in Change" Project Survey, 2012.

2.2.4 Interviews with 60 journalists

The purpose of the interviews was to get deeper into the areas also covered by the survey. Twenty interviews were accomplished by the researchers or by assistants in each country, and they were based on the same interview guide (see Appendix 2). The interviews were mostly done face-to-face, but in some long distance cases in Russia also on Skype. Usually each interview lasted about an hour.

The following areas were covered during the interview:

- Professional history – experience, training, why become a journalist.
- Professional practice – daily work, multiskilling position.
- Professional values.
- Professional autonomy – pressure from inside and outside, limits for what is possible.
- Relationship to the audience.
- Future – for themselves and for the profession.

The selection of journalists for the interviews was done to achieve a great variety in age, gender, types of media and geographical position in the country. The journalists were chosen in a "snowball selection" with personal contacts in the beginning of the chain (a list of all respondents is attached in the end of the book, country by country). All respondents are identified with age, gender, and position in the newsroom, as well as type of media and type of place in the country. Each respondent is given a signature, for example Swedish respondent no. 1 = SR1, Polish respondent no. 7 = PR7, and Russian respondent no, 14 = RR14. In that way it is possible for the reader to identify what kind of answers and opinions belong to the same respondent.

All interviews were taped, transcribed, and finally translated into English in order to be available for all researchers in the project.

2.2.5 Comparative analysis

Comparative analysis is one of the most important methods in social and communication investigation nowadays. This approach has been very useful also in the project "Journalism in Change." We used a tri-nation comparison method in each aspect of our research where contextual interrogation precedes any analysis of similarity and difference (see Hallin and Mancini, 2004:2–3). This method let us test the hypotheses presented at the beginning of our research (see Figure 2.2).

Figure 2.2: Relationship of comparative methodological choices in the project "Journalism in Change"

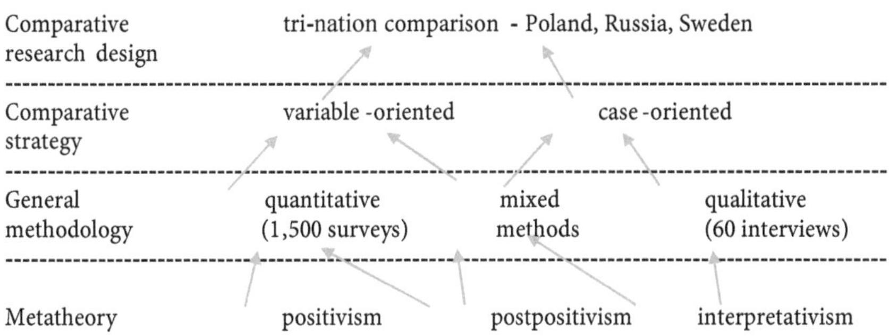

Source: Authors' adaptation on the base of Lor (2011:5)

Thanks to the data elaborated in quantitative research (1,500 surveys), we can map many combinations of variables observed in the data set. We are able to operationalize the variables by using a set of indicators such as numbers, per cent, means, standard deviation, Pearson correlation, and factor analysis in order to compare them in each aspect of our research. The findings are reported and illustrated in tables and figures. The qualitative comparative research is an important part of this study, and it enriches the quantitative one. In addition, data gathered from 60 in-depth interviews allows us to discover and compare some aspects of the journalistic cultures in each country, and to make many observations.

Thanks to comparative analysis it is easier to find similarities and differences among Polish, Russian and Swedish journalists, and to better understand the changes in journalism.

2.3 Summary

There are many potential problems in comparative journalism research. One of the major challenges is the danger of carrying western bias in research of journalism outside the liberal Anglo-Saxon world. Questions and models used to analyze journalism in western countries cannot easily and directly be transferred to comparative research covering countries outside liberal western countries. Notions like "objectivity" and "integrity" might not carry the same meaning in different professional cultures.

To handle these problems, the research project "Journalism in Change" uses a design based on multinational cooperation and a combination of methods. In the team, researchers from Södertörn University in Sweden have cooperated with researchers from the University of Wrocław in Poland and Moscow State University in Russia. The cooperation has been ongoing through the whole project; from the planning of the project, the construction of the survey and interview guide, accomplishing the empirical studies with national teams, to the final analysis in writing of reports, articles and this book. The cooperation has been very tight and the result is as free from nationally rooted bias as possible.

With a combination of methods it is possible to go deeper than in other comparative journalism research. The survey gives a general picture of areas covered in the project, and in the interviews journalists get space to explain and reflect on the questions raised in the project. Together the survey and the interviews produce a broad and reliable picture of the professional cultures in the three countries.

3. Who is a journalist today? Mapping selected dimensions for comparative study on journalism

Michał Głowacki

3.1 Introducing selected dimensions for studies on journalism

Hundreds (or probably even thousands) of different studies, reports and analyses on journalism have been conducted and presented around the world. However, answering the question of "Who is a journalist?" has always been difficult and has become even more complex and fragmented in the digital age, with new media and technologies and network practices. Academic works, legal cases and representatives of the media industry have developed several different approaches either with a reference to professional status, culture, values and beliefs, ideals and practices or by referring to the role of journalists in society (Hanitzsch, 2007, Fengler et al., 2014). In addition some of the most recent discussions have just offered room for the analysis of changing patterns in journalism (Zelizer, 2009), professional identity (Witschge and Nygren, 2009), the role of culture (Deuze, 2007) as well as examples of professionalization (Waisbord, 2013), "de-industrialization of information," and "de-ritualization of news consumptions" (Broersma and Peters, 2013). The rise of media activism, collaborative media-making, new innovation regimes and convergence culture offer several new approaches. In the times of new notions of media, including media-like services, such as blogs, connective platforms (Facebook, Twitter), wikis, mobile devices and so on, traditional attempts to define journalists continue to be questioned and/or updated.

Defining journalists has never been an easy task. For instance, when Tunstall (1971) attempted to map features of 'specialist newsgathering journalists' in the United Kingdom, he mostly looked at dimensions related to occupation, journalistic goals and new routines. Bohere (1984) put the emphasis on ethical standards, functions in the newsrooms, journalism training, career opportunities, industrial relations as well as issues related to safety and social security of employees. The analysis conducted within the international research project "Mediadem" project in fourteen European countries further underlined the existence of two models: status- based definition and activity-based definition.

The data also underlined the lack of a universal approach, since in some countries journalists have been defined by law and in some other countries by private regulation (Cafaggi et al., 2012). When in October 2013 a research team coordinated by Peters (University of Dayton) and Tandoc Jr. (Missouri School of Journalism) in the U.S. attempted to analyze a variety of sources that conceptualize journalism, they ended up with a descriptive working definition of journalist, as

> "(...) someone employed to regularly engage in gathering, processing, and disseminating news and information to serve the public interest." (*Who is a journalist?*, 2014)

Although they believed that definition with the focus on activities, output, and the social role could point out features of journalism in the digital era, they also noted that

> " (...) by referring to employment, however, the definition delivers a fatal blow to the people engaging in many new forms of journalism." (Peters, 2013: 62)

In one of the comments related to the study outcomes, Peters (2013) further elaborated on why "defining a journalist is messy, but crucial." The need for developing a definition of journalists is of high importance in case law as well as in every scientific research project, which aims at analyzing journalism either from the national or comparative perspective. Although it would be interesting to get involved in the discussions about the nature, complexity and future approaches to journalism, this is not our goal here. By posing the question "Who is a journalist today?" we aim here at offering a scheme for journalism cross-national analysis, which serves as a reference point and a background for explanation of the findings presented in this collection. With regard to the results from "Journalism in Change", empirical studies as well as findings and insights from in-depth interviews with media practitioners in media outlets, we look at selected dimensions in order to draw a picture of journalism in Poland, Russia and Sweden. Bearing in mind different traditions and national perspectives, sample strategies, as well as the complexity of approaches in changing media landscapes, we map selected dimensions for study on journalism (see Figure 3.1).

Figure 3.1: Mapping selected dimensions for study on journalists

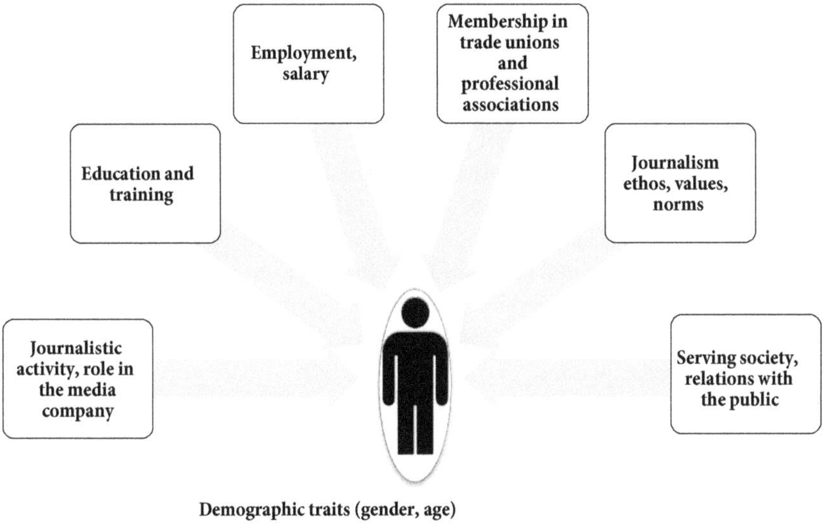

Source: Author.

The analyses presented in this contribution focus on:

a) education and training: level of education and the role of journalism training;
b) employment and salary: permanent vs. temporary employment, freelance journalism, full-time vs. part time;
c) membership of trade unions and professional associations[1].

However, one should note that there are strong interrelations between the proposed dimensions. For instance, to fully understand types of employment and working relations in newsrooms, issues related to journalistic activities, daily practices as well as functions and the roles in the newsroom need to be taken into account. Moreover, journalism practices are deeply rooted in values and norms as well as journalism and organizational culture. All of this when taking into account selected demographic traits – gender and age, which are analyzed

1 Data presented here does not take into account the notion of journalism ethos, since values, norms, and beliefs are widely discussed by Maria Anikina in Chapter 6. For a detailed explanation of journalistic activities, multitasking and daily journalistic work, see Chapter 4 by Jöran Hök and Chapter 5 by Gunnar Nygren.

here in connection to each selected dimension[2] and are further discussed in other chapters of this collection.

The salient questions we address are: What is the gender and age proportion in each country? What is the level of education of journalists in each country? How many of the respondents have obtained their degree in journalism? What are the employment conditions of journalists? Do most of them work full-time and carry out the permanent position in a media organization? How do journalists surveyed perceive the importance of salary when choosing a profession and getting involved in journalistic activity? How many of the journalists surveyed belong to professional association(s) and/or trade union(s)? What are the similarities and differences in the perception of journalism in Poland, Russia and Sweden as well as between different age and gender groups in each country? And how might they be explained?

3.2 Gender and age proportions

Both in Russia and Sweden more women than men indicated working as a journalist (see Figure 3.2). The study sample mapped 62 per cent of female journalists in Russia, which further supported the previous study on Russian journalism by Pasti et al. (2012)[3]. Similar to this, 53 per cent of women in the "Journalism in Change" study have confirmed a small majority of female journalists in Sweden, which in the previous study was defined by Strömbäck, Nord and Shehata (2012) at the level of 50 per cent. Although the number of Polish women working as a journalist has grown in recent years, the profession is still dominated by men. Data derived from both "Journalism in Change" indicated 58 per cent of male respondents in a country.

Although it is hard to explain why Polish journalism has been traditionally dominated by men,male interviewees in Poland claim there are no significant barriers and gender differences; as one of them noted:

> "I think they [women – M.G.] are more sensitive, and it has its advantages (…). Men are rather tough and especially journalists are generally considered to be arrogant, crafty, so on, which is of course true. (…) The more women in the profession, the better." [PR1, male, under 35 years old, radio, reporter, editor-in-chief, new media department]

---

2 Due to space limitations this Chapter does not deal with types of media (press, broadcasting, new media) and positions in the newsrooms (managerial vs. operational level) as variables for comparative studies.

3 In "The Global Journalists in the 21st Century" Pasti et al. (2012) indicated 60 per cent of female journalists in Russia.

Figure 3.2: Gender of respondents (in per cent)

Source: *Your gender?* The "Journalism in Change" Project Survey, 2012.

In both Russia (69 per cent) and Poland (42 per cent) the youngest generation of journalists (under 35 years old) created the biggest professional group (see Figure 3.3). Only 26 per cent of Swedish respondents have been mapped in this age category. The data might be explained by a generation shift in Russia and Poland in connection to social and political transformations in the late 1980s and early 1990s. In Sweden the profession is much older with the largest group over 50 years (38 per cent). The profession has grown older, in 1989 only 19 per cent of journalists were 50 years and older. This shows the generational shift in Sweden was earlier than in Poland and Russia, probably during the fast expansion of journalists in the 1970s and1980s (Nygren, 2008).

Figure 3.3: Three age groups of respondents (in per cent)

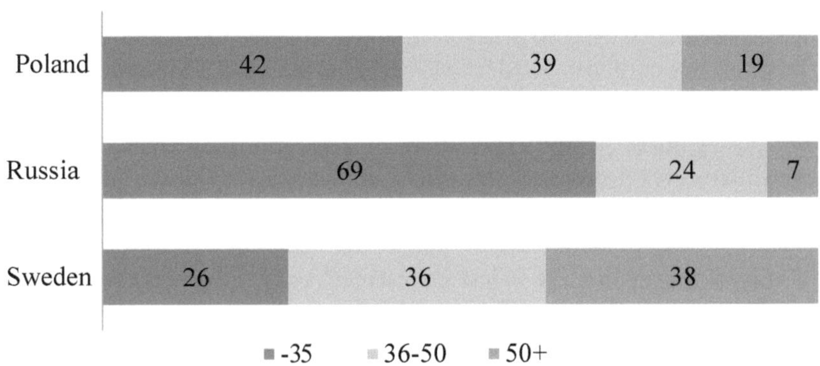

Source: *What year were you born?* The "Journalism in Change" Project Survey, 2012.

There are more men than women in each of the age groups in Poland, although differences in the youngest generation are not that significant (54 per cent of men and 46 per cent of women in the age group under 35). The data confirms that the majority of Russian journalists surveyed (with the exception of the age group 36–50 years old) are women. The biggest group of women in Russia (71 per cent) and Sweden (62 per cent) belong to the youngest generation of journalists (See Table 3.1). The data in these two countries might reflect the fact that more women than men attend professional journalism education.

Table 3.1: Gender in each of the age categories

	Under 35 years old				36–50 years old				Over 50 years old			
	Female		Male		Female		Male		Female		Male	
	N	%	N	%	N	%	N	%	N	%	N	%
Poland (N 460 = 100 %)	89	46	106	54	74	41	107	59	30	36	54	64
Russia (N 500 = 100 %)	246	71	101	29	47	40	70	60	19	53	17	47
Sweden (N 443 = 100 %)	78	62	48	38	96	57	74	42	74	42	104	58

Source: The "Journalism in Change" Project Survey, 2012.

Some of the interviewees in Russia have indicated to what extent the factor of age might have on the quality and the nature of journalism in the country. Although most young people do not see many differences between attitudes of journalistic generations one of the youngest Russian media professionals claims that:

> "Young journalists today are all aggressive, insolent, fast, but they constantly make slips of the tongue, do not watch the quality of their speech. The "older" generation of journalists didn't make such missteps, and I stand with them in this regard." [RR4, male, 25 years old, sports observer of federal radio]

When posing questions and explanations on generational differences, the older representatives of the profession have not completely agreed on the universal image of young people, finding the biggest differences in the usage of new technologies and content production online. One of the studies by Erzikova and Lowrey (2011) have shown that the oldest representatives of the profession in Russia have a negative view of their colleagues; they see the younger generation as being "disconnected" from their own country/region, unskilled, poorly educated and motivated and rather submissive to authority. Similarly, several negative assumptions have also been noted in Poland, where journalists criticized their younger colleagues mostly for the lack of skills as well as attitude of young people when entering the profession:

> "I've noticed that the moment they come here, on their first day, they already want to appear in front of the microphone, they want to host a live program because they think they are super great etc., wonderful, right? Great that young people are so brave, but kind of they don't seem to realize that they should learn more beforehand." [PR9, female, 35 years old, private radio]

A similar picture has been drawn by one of the Swedish journalists, who claims that:

> "Trainees have neither the feeling nor the humility to confess that "I'm not so good so I'll have to ask an older colleague" (…). I myself am so incredibly afraid of doing something wrong, so I have to ask. Maybe that's the feeling that they lack." [SR18, male, 41 years old, web editor magazine]

By contrast, older generations of journalists have been seen by young journalists as being more idealistic and pragmatic. Overall, age differences have also been explained by general trends in the newsrooms, namely as changes from specialized positions towards general reporting, increased multiskilling, convergence strategies, and so on (Nikunen, 2013).

Gender and age groups as well as journalistic generations create an interesting dimension when explaining the attitudes of journalists towards new technologies, since a large number of the youngest representatives of the profession grew up in the rise of the Internet era and have developed different patterns of media use, including connective (social) media (see Chapter 9). Moreover, in Poland and Russia age categories might be used to examine the perception of journalism roles as well as media-political relations among the youngest generation and their older colleagues, who either entered the profession in the time of social and political transformation, or started in the period of communism. Finally, this might be helpful when investigating changes in journalists' expectations and perception of roles in society. From this perspective it will be interesting to find out whether gender and age proportions make a difference in journalism cultures, perception of ethical values (see Chapter 6). For instance, as argued in one of the previous studies on generations of journalists in Russia:

> "The crucial difference between the new generation and the old is the idea of journalism as a marketplace, competition, race and battlefield where there are no ethical norms and corporate loyalties." (Pasti, 2005: 109)

3.3 Education and training

Both type and level of education chosen by each individual have a significant impact on his/her future career path. Overall, no universal model for education has been and can be applied; education should be understood in a broader sense, including college and university degrees on the one hand as well as training and practical skills on the other. In addition to this, education is a life-long learning process rooted in individual reasons and perception, starting from motivation, commitment, curiosity, interests, family tradition, prestige, expectations and/or a need for personal development.

In most European countries, including Poland, Russia and Sweden, education in media and journalism is not a prerequisite for entering the profession. However, in some media organizations professional background, training, and/or internship might be necessary or/and helpful when applying for a job. Bearing in mind different traditions and education systems, one could ask: What kind of degree do journalists have? What kind of knowledge and experience do they gain from training courses and classes offered by schools and universities? And to what extent does education in journalism contribute to professionalization, perception of values and the quality of journalism profession in a given country? Some research has been made in recent years on journalism education and relation to the industry in the three countries (Nygren et al., 2010; Stigbrand and Nygren, 2013). In this study we analyze education from the other perspective – the perspective of active journalists.

There is a significant majority of professionals with a higher education degree in each country (Figure 3.4). In Russia, 96 per cent of journalists surveyed graduated from university (higher education) (more than 90 per cent of both women and men). This tendency is also relevant when looking at the level of education in relation to age groups; all of this suggests that a need for having a degree in Russia is extremely important and attracts attention. In Sweden, where the high level of education dominates (90 per cent), there are also more women than men (93 per cent vs. 85 per cent) with this type of degree. Higher education in this country is more often obtained by the youngest representatives of the profession. An explanation for this could be related to the fact that most of the older journalists entered the profession before the competition of the media market was high. Poland is the only country where the percentage of young people surveyed with higher education (84 per cent in the group under 35 years old) is lower than in other age groups (91 per cent in 36–50 years old and 97 per cent in over 50 accordingly) (see Table 3.2). Men being reported as the majority of journalists are less educated than women (see Table 3.3).

Figure 3.4: Level of education of journalists (in per cent)

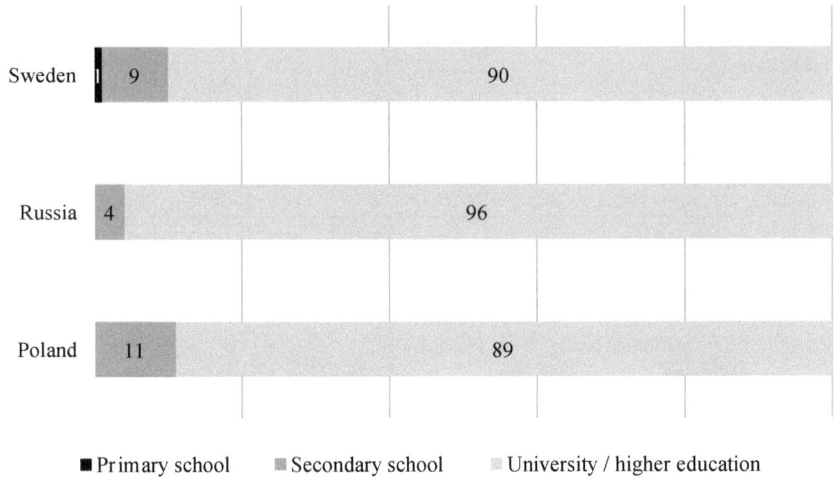

Source: The "Journalism in Change" Project Survey, 2012.

Table 3.2: Level of education with respect to age

Education		Primary school		Secondary school		University/higher education	
		N	%	N	%	N	%
Poland	Under 35 (N 191 = 100%)	0	0	30	16	161	84
	36–50 (N 176 = 100 %)	0	0	16	9	160	91
	Over 50 (N 83 = 100 %)	1	1	2	2	80	97
Russia	Under 35 (N 347 = 100 %)	0	0	16	5	331	95
	36–50 (N 115 = 100 %)	0	0	3	3	112	96
	Over 50 (N 36 = 100 %)	0	0	1	3	35	97
Sweden	Under 35 (N 125 = 100 %)	0	0	5	4	120	96
	36–50 (N 169 = 100 %)	0	0	17	10	152	90
	Over 50 (N 166 = 100 %)	4	2	23	13	150	85

Source: The "Journalism in Change" Project Survey, 2012.

Table 3.3: Level of education with respect to gender

Education		Primary school		Secondary school		University/ higher education	
		N	%	N	%	N	%
Poland	Men (N 279 = 100 %)	1	0	43	16	235	84
	Women (N 204 = 100 %)	0	0	9	4	195	96
Russia	Men (N 187 = 100 %)	0	0	14	8	173	92
	Women (N 311 = 100 %)	0	0	6	2	305	98
Sweden	Men (N 234 = 100 %)	4	2	30	13	200	85
	Women (N 259 = 100 %)	0	0	17	7	242	93

Source: The "Journalism in Change" Project Survey, 2012.

When searching for comments and explanations on the data presented above, one should take into account the quality of education and knowledge that is being offered at every stage of the education process. For instance, some of the interviewees in Poland questioned the general knowledge and skills of people who recently graduated from universities. Two opinions illustrate this issue more specifically:

> "Surely these are people [young journalists – MG] who have less general knowledge and erudition than journalists of my generation." [PR17, male, 35–50 years old, national daily newspaper]

> "Nowadays generally people who graduate from universities, from schools have less knowledge. Maybe they are better educated in one narrow highly-specialized area but they definitely have less general knowledge." [PR10, female, over 50 years old, local daily newspaper]

What is the role of journalism education then? Much could be written about academic institutions offering classes and courses for future potential journalists in each country. For instance, in Russia "journalism is classed among the humanities, and one of the most important tasks of the media is to shape the cultural needs of society" (Vartanova et al., 2010: 207). In Poland future media professionals who want to be trained in journalism might obtain a degree from eighteen public universities and thirty-five private schools, which usually offer courses on journalism and communication either within the faculties of linguistic studies or political and social science (Szot, 2009; Dobek-Ostrowska, 2011). In Sweden journalism studies have been traditionally connected to universities but also vocational education at a lower level. In recent years the education market has grown rapidly and had an impact on changes in the structure of the profession. As argued by Ghersetti and Weibull:

> "Expansion of academic education has been a very important means for the increase in the number of female journalists." (2009: 201)

Looking at the study results one might learn that a significant percentage of respondents in both Sweden (55 per cent)[4] and Russia (75 per cent) have obtained their degrees in journalism from universities (see Figure 3.5). Moreover, in both Russia and Sweden the majority of professionals holding a journalism degree at an academic level represent the younger generation (82 per cent and 70 per cent respectively) and – at the same time – there is rather little importance of training in media companies (see Table 3.4). Although the 'gap' between women and men who obtain a degree in journalism is bigger in Russia than in Sweden, the role of journalism education in both countries is significant and might be explained in several different ways (see Table 3.5).

One could argue that in Russia and Sweden studying journalism is prestigious and offers great opportunities for finding a job in the media industry after graduation. In addition to this, as earlier research in both countries confirmed, students of journalism mostly chose the profession because they found it interesting and they want to develop their own potential (Nygren et al., 2010). On the other hand, this might be explained by the expansion of education institutions, growing competition in the job market, even family traditions or simply a personal need to study journalism before entering the profession. Changes in the education market and media industry in Sweden are pointed out by one of the interviewees:

> "(...) twenty years ago (...) only 40 per cent of those who worked there as journalists were educated for it. Nowadays, I think it's hard to get a job without the suitable education. If you are young today and want to get a job in media, you need education in journalism. That is a difference, that today one must have an education to get a chance of a job, it wasn't like that before." [SR20, male, 41 years old, web editor]

On the other hand, in Russia some of the interviewees undermine its role in the context of quality and skills being offered by schools of journalism:

4 In "The Global Journalists in the 21st Century" Strömbäck et al. (2012) indicated 78 per cent Swedish journalists with a University degree and 46 per cent of journalism graduates.

"I have no professional education in journalism, no way (…) I am convinced that a major in journalism is not a prerequisite for working as a journalist." [RR15, 36–50 years old, male, national weekly magazine]

"Journalism departments no longer teach the profession, but give a general universal education, with which you can go to different areas." [RR19, female, 50 years old, national newspaper]

About 43 per cent of journalists in Poland do not hold a degree in journalism (Figure 3.5). This number is bigger than the percentage of professionals obtaining a journalism degree at the academic level; 40 per cent of journalists with a journalism degree from academia places Poland in opposition to tendencies in Russia and Sweden. The other distinctive feature is the proportion of trainees in Polish media enterprises, which is the highest among all the countries compared. At the same time, similar to Russia and Sweden, an academic degree in journalism is the most popular among representatives of the youngest generations and women (see Table 3.4).

Figure 3.5: Journalism education of journalists (in per cent)

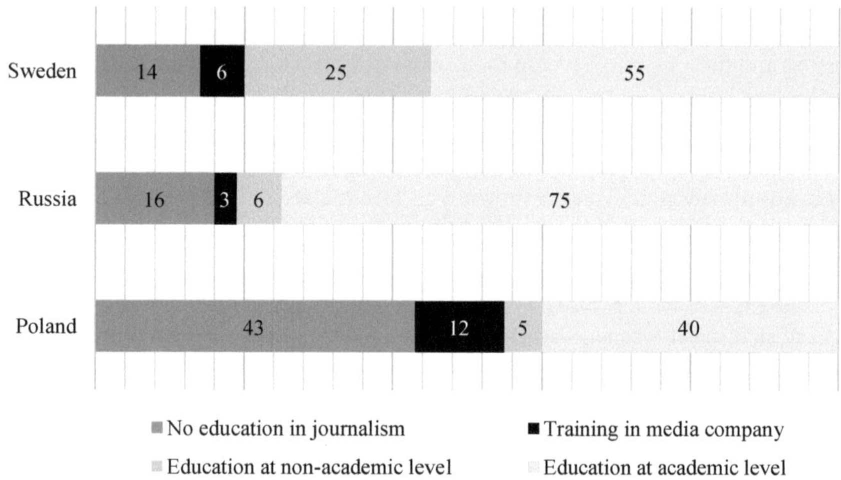

Source: *Do you have a special education in journalism?* The "Journalism in Change" Project Survey, 2012.

Table 3.4: Journalism education with respect to age

Education in journalism		No journalism education		Training in media company		Education at non-academic level		Education at academic level	
		N	%	N	%	N	%	N	%
Poland	Under 35 (N 180 = 100 %)	74	41	13	7	13	7	80	45
	36–50 (N 173 = 100 %)	84	48	22	13	7	4	60	35
	Over 50 (N 81 = 100 %)	33	41	13	16	1	1	34	42
Russia	Under 35 (N 336 = 100 %)	38	11	12	3	13	4	273	82
	36–50 (N 116 = 100 %)	33	28	2	2	10	9	71	61
	Over 50 (N 36 = 100 %)	8	22	2	6	7	19	19	53
Sweden	Under 35 (N 125 = 100 %)	12	10	2	2	23	18	88	70
	36–50 (N 167 = 100 %)	25	15	7	4	47	29	88	52
	Over 50 (N 176 = 100 %)	28	16	17	10	46	26	85	48

Source: *Do you have a special education in journalism?* The "Journalism in Change" Project Survey, 2012.

Table 3.5: Journalism education with respect to gender

Education in journalism:		No journalism education		Training in media company		Education at non-academic level		Education at academic level	
		N	%	N	%	N	%	N	%
Poland	Men (N 273 = 100 %)	129	47	33	12	14	5	97	36
	Women (N 193 = 100 %)	73	38	21	11	8	4	91	47
Russia	Men (N 187 = 100 %)	36	19	8	5	23	12	120	64
	Women (N 311 = 100 %)	43	14	8	3	7	2	253	81
Sweden	Men (N 232 = 100 %)	40	17	16	7	55	24	121	52
	Women (N 259 = 100 %)	29	11	11	4	69	27	150	58

Source: *Do you have a special education in journalism?* The "Journalism in Change" Project Survey, 2012.

Journalism in Poland is taught in prestigious public universities; after the social and political transformation of 1989 the educational offered in journalism has been extended to private colleges and schools. Still, courses are among the most popular in the group of social science and hundreds of new students begin their journalism education every academic year. Thus, in taking a closer look at the

Polish case, as well as previous studies underlining similar patterns[5], it is essential to elaborate what are the reasons responsible for limits of the potential of journalism education?

The research conducted in the international project "Media Accountability and Transparency in Europe" (MediaAcT)[6] indicated that 54 per cent of media professionals in Poland think that journalists in a country are inadequately trained. Interestingly enough, the problems with education were more often observed by professionals representing managerial level (66 per cent of respondents) and journalists from older generations (Głowacki, 2013). Naturally, one shall look for an explanation history of studying journalism. As argued by Hiebert and Gross in the first post-communist years:

> "(…) journalism educators, with few, qualified exceptions, had no starting point or background for radically transforming professional education/training in their classrooms." (2003: 263)

Among the other problems facing the education system in Central and Eastern Europe the authors point out the lack of sufficient funding for journalism education, poor professors' salaries and "in programs that have been ongoing for a long time" (Hiebert and Gross, 2003: 264). This varies a lot starting from courses on the history of media, theory of communication, media ethics, public relations, advertising and so on. Several universities in a country offer practical workshops and even an opportunity to get involved in university-based community media, such as students' radio, magazines and websites. The questions could be "in what way doesthe Russian model differ from the Polish model of journalism education? What kind of skills do Polish journalism graduates possess and how are practical workshops at the universities and the role of academia being recognized by representatives of the industry?" One of the interviewees in Poland underlines the need for practical education:

---

5 In "The Global Journalists in the 21st Century" Stępińska et al. (2012) indicated 84 per cent of Polish journalists with a University degree and 32 per cent as journalism graduates.

6 The EU funded project "Media Accountability and Transparency (MediaAcT)" conducted in 2010–2013, aimed at investigating media accountability in the digital era in twelve European countries: Austria, Estonia, Finland, France, Germany, Italy, The Netherlands, Poland, Romania, Spain, Switzerland and United Kingdom as well as two countries in the Arab world (Jordan and Tunisia). For information and the outcomes of the project see www.mediaact.eu.

> "A young person studying journalism but not practicing it – it's an example of what the whole generation does, or just a misunderstanding. It's not a profession that you can just learn." [PR13, 36–50 years old, female, freelancer]

While two others point out that the profession is open to everybody, and that knowledge and skills from fields other than journalism might be more appreciated:

> "(...) when a journalist has a degree in some other field, different from journalism, it means that they are good at other subjects (...), education in a different field is even an asset here." [PR2, female, 30 years old, regional branch of national newspaper]

> "You can be a journalist by profession, not by the degree that you have (...). When I look for a journalist, I check if he has any knowledge about the world in general." [PR5, female, 36–50 years old, public service TV]

Drawing the full picture of the role of journalism education in Poland would require additional research, examining the impact of academia on journalism behavior, relations between scholarly institutions and the media industry, recruitment requirements adopted by media enterprises, differences in importance between university skills and professional training in given media outlets as well as career paths of young people who do not find a job in the media after graduating from a school of journalism. In relation to all the countries analyzed here it will also be important to observe whether differences in the education of journalists in all the countries change the professional values of the actual news reporting. Content analysis could further indicate to what extent differences in perception of journalism education contribute to the quality and level of journalism professionalization. As researchers from one of the most recent studies on media accountability and responsiveness indicate that:

> "Respondents not trained in professional ethics are less supportive of new media – based accountability instruments, such as interaction with users." (Radu and Popa, 2014: 260)

Thus, the analysis presented in this section might help to understand in-depth analyses of values, norms and relations with media users presented in Chapter 6 and in this collection.

3.4 Employment and salary

As argued in the previous sections, there might be a plethora of explanations why individuals decide to choose the profession. This might vary a lot; the list of the most popular reasons includes serving society, motivation, family traditions, willingness to become famous and recognized, job market opportunities as well as expectations related to salary or future professional experience. In addition, one will further map several different factors that play a role when recruiting a new person to a job or building and developing teams in the newsrooms. In this vein the analysis of working conditions and employment should focus on recruitment processes adopted by a given media enterprise, job design, human relations, organizational structures, span of control, decision-making processes, mechanisms supporting staff development and promotion system at work, vision and strategy of the organization and so on. Bearing this in mind the complex picture, we focus here only on selected dimensions that might be helpful to indicate the main similarities and differences of journalism in Poland, Russia and Sweden. All the data shall be further discussed in relation to general knowledge on structure and financial conditions of each media market as well as the current situation of national economy and employment.

Starting in 2008 the global financial crisis/slowdown has had a profound effect on national economies and hundreds of industries, including media. On the other hand the processes of media convergence as well as the rise and development of media-like services have dramatically changed the structure of media landscapes globally. A shrinking work force and decreasing number of people in the newsrooms has resulted from crises in the business models in traditional media companies (most notably in print media) and which have also been observed in Poland, Russia and Sweden. Several other trends, including changes in production, multiskilling and forms of employment contribute to a broader picture of journalism in all three countries today (see Chapter 4).

When looking at the structure of the profession, regular (permanent) employment in media companies seems the common feature in Poland, Russia, and Sweden; its level varies from 67, 74 and 72 per cent of respondents accordingly (see Figure 3.6). All the countries analyzed in the study indicated similar tendencies with regard to the proportion of temporary employment and freelance journalists. Additional findings appear when one takes into account differences in relation to age groups and gender. Both in Poland and Sweden only around 50 per cent of journalists from the youngest group have the status of regular employee in a given media company. In Russia, by contrast, more than 70 per cent of journalists surveyed from the under 35 years old age group indicated having permanent employment at the media company (see Table 3.6). It is remarkable that journalists from the biggest age group in a country (see Figure 3.3) enjoy all the benefits, including social security, stability at the job market etc. In all the countries examined more men than women have the status of regular employment. In Russia and Sweden – where women are in the majority and are better educated – gender dissemination shall be therefore analyzed in line with employment policies (see Table 3.7).

Figure 3.6: Types of employment (in per cent)

Source: *What kind of main employment do you have today (one option)?* The "Journalism in Change" Project Survey, 2012.

Table 3.6: Types of employment with respect to age

Types of employment		Regular		Temporary		Job agency		Freelance		Other	
		N	%	N	%	N	%	N	%	N	%
Poland	Under 35 (N 194 =100%)	95	49	40	21	1	0	28	14	30	16
	36–50 (N 180 = 100 %)	146	81	7	4	0	0	18	10	9	5
	Over 50 (N 82 = 100 %)	63	77	3	4	0	0	10	12	6	7
Russia	Under 35 (N 346 =100%)	246	71	35	10	1	0	62	18	2	1
	36–50 (N 117 = 100 %)	91	78	11	9	0	0	15	13	0	0
	Over 50 (N 36 = 100 %)	27	75	8	22	0	0	0	0	1	3
Sweden	Under 35(N126 = 100%)	62	49	33	26	7	6	19	15	5	4
	36–50 (N 170 = 100 %)	134	79	10	6	4	2	22	13	0	0
	Over 50 (N 178 = 100 %)	144	81	5	3	0	0	29	16	0	0

Source: *What kind of main employment do you have today (one option)?* The "Journalism in Change" Project Survey, 2012.

Table 3.7: Types of employment with respect to gender

Types of employment		Regular		Temporary		Job agency		Freelance		Other	
		N	%	N	%	N	%	N	%	N	%
Poland	Men (N 286 = 100 %)	205	72	25	8	0	0	33	12	23	8
	Women (N 202 = 100 %)	121	60	28	14	1	0	28	14	24	12
Russia	Men (N 187 = 100 %)	145	77	16	9	0	0	25	13	1	1
	Women (N 311 = 100 %)	219	70	38	12	1	0	52	17	1	0
Sweden	Men (N 234 = 100 %)	175	75	20	9	5	2	33	14	1	0
	Women (N 262 = 100 %)	180	69	30	11	7	3	41	16	4	1

Source: *What kind of main employment do you have today (one option)?* The "Journalism in Change" Project Survey, 2012.

In order to draw a more complete picture of employment structure in each country one can consider how types of employment are reflected in relation to full- and part time activities (see Figure 3.7). The biggest proportion of journalists working full-time might be found in Sweden (93 per cent), where 72 per cent of employees indicated having a permanent job at a media organization. One needs to underline that in Sweden there are no significant differences among the gender and age groups (Table 3.8 and Table 3.9). This might suggest that even a person who has a temporary contract at a media organization or work as a freelance, invest most of his/her time devoted to work into journalistic activities. This type of attitude might require entrepreneurial skills, the ability to work on different projects at the same time as well as a high level of flexibility in the job market, which is especially relevant to those without a permanent post at a media enterprise. An example of this type of attitude is presented by one of the youngest journalists in Sweden:

> "I combine freelance and study so I can really work as much as I want. If I run a very hard race, I have a plenty of jobs." [SR16, female, 28 years old, freelancer, reporter at the newspaper]

Figure 3.7: Full-time vs. part-time activities (in per cent)

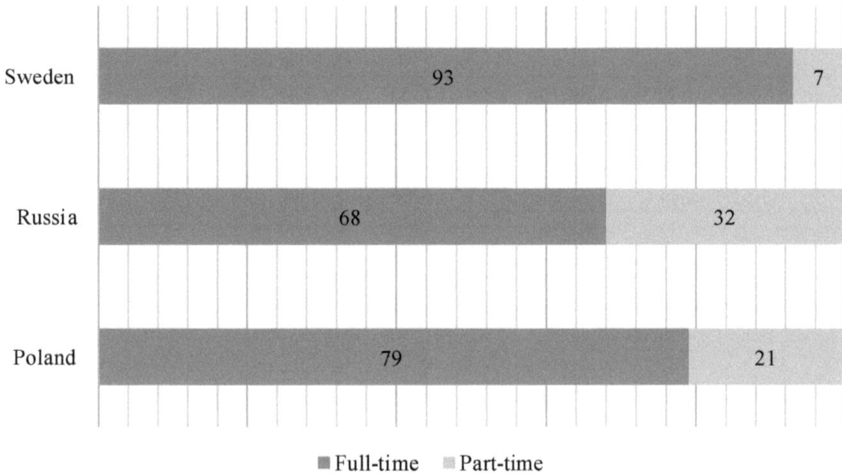

Source: *How many hours are you working as a journalist in an average week?* The "Journalism in Change" Project Survey, 2012.

Table 3.8: Full-time and part-time activities with respect to age

Working hours		Full-time activities		Part-time activities	
		N	%	N	%
Poland	Under 35 (N 190 = 100 %)	127	67	63	33
	36–50 (N 175 = 100 %)	157	90	18	10
	Over 50 (N 80 = 100 %)	68	85	12	15
Russia	Under 35 (N 346 = 100 %)	225	65	121	35
	36–50 (N 115 = 100 %)	90	78	25	22
	Over 50 (N 36 = 100 %)	25	69	11	31
Sweden	Under 35 (N 126 = 100 %)	119	94	7	6
	36–50 (N 170 = 100 %)	160	94	10	6
	Over 50 (N 176 = 100 %)	160	91	16	9

Source: *How many hours are you working as a journalist in an average week?* The "Journalism in Change" Project Survey, 2012.

Table 3.9: Full-time and part-time activities with respect to gender

Working hours		Full-time activities		Part-time activities	
		N	%	N	%
Poland	Men (N 282 = 100 %)	233	83	49	17
	Women (N 194 = 100 %)	145	75	49	25
Russia	Men (N 186 = 100 %)	52	28	134	72
	Women (N 311 = 100 %)	206	66	105	34
Sweden	Men (N 234 = 100 %)	220	94	14	6
	Women (N 260 = 100 %)	238	92	22	8

Source: *How many hours are you working as a journalist in an average week?* The "Journalism in Change" Project Survey, 2012.

In Poland, where a total of 79 per cent of media professionals indicate working full-time, differences between gender groups are more significant. Above all, the data indicate that the situation of the youngest representatives of the profession on the job market is much more difficult (49 per cent of respondents with regular employment and 67 per cent working as journalists full time). This might suggest that the potential of young people in Poland might not be fully recognized by employees. At the same time, since the data relating to young people with regular employment is comparable to the Swedish case, one might ask what role is played by journalism education, young people's expectations, skills, experience and/or flexibility of young people in the job market. These results could also suggest that young people in Poland need to combine journalistic with non-journalistic activities. Here the questions are: How difficult is it to get a job in journalism (including freelancers) and what is the impact of the financial slowdown and/or crisis to the recruitment processes in media enterprises? Several thoughts gathered during the interviews make attempts at explaining some of these tendencies:

> "(…) young people don't realize that when they begin their studies, they don't check the labour market, and they don't know if it is easy to find a job in the profession." [PR10, female, more than 50 years old, local newspaper]

> "(…) the journalism industry is in crisis right now and you can hear about people being made redundant or about pay cuts." [PR15, male, under 35 years old, private radio]

In Russia, where 74 per cent of journalists have a permanent post at a media outlet, 68 per cent of media professionals work full time. Similar to other countries, there are differences with respect to gender groups. In addition, 65 per cent of the youngest people work as a journalist full-time. Bearing in mind the three age groups defined in one of the previous sections (see Figure 3.2) as well as data in relation to types of employment (see Figure 3.6) one might ask why

35 per cent of the youngest people do not invest the whole amount of professional time to work as a journalist. In searching for an answer one should more specifically look at organizational structures, market opportunities as well as capabilities of media outlets to employ young professionals full time. In line with the previous studies indicating that many journalists in Russia hold down a second job (Pasti et al., 2012), it will be worth exploring what jobs they have and to what extent getting involved in additional work activity depends on salary. The ideas gathered during the interviews in Russia offer some preliminary insights in this respect:

> "I know some journalists who decided to go "cross-functional", working at several media outlets and in different areas of journalism in order to make more money, repay a loan or purchase something expensive." [RR4, male, 25 years old, sport observed]

> "(…) [being a freelancer – MG] does not bring money — only the satisfaction that your materials are wanted by the readers." [RR3, female, 62 years old, freelancer, newspaper]

Overall, in Poland and Russia media professionals agreed that salary plays an important role when working as a journalist. The comparison of research means indicated a relatively high level of agreement in both countries, while the importance of income in Sweden was lower[7]. In taking a closer look at the role of salary when choosing the profession, journalistic perception is again the biggest in the Polish and Russian case (study means 4,10 and 4.08 accordingly). In both countries more attention when entering journalism was indicated to salary than to job security and influencing society. On the contrary, Swedish journalists indicated that security at work plays a more important role than income when choosing career in journalism (study means 3.89). Serving the public interest as reason for choosing the profession gathered more attention than salary in Poland and Sweden (see Table 3.10).

Table 3.10: Reasons for choosing the profession

	Salary			Job security			Benefits			Serving the public interest			Influencing society		
	N	M	SD	N	M	SD	N	M	SD	N	M	SD	N	M	SD
Poland	474	4.10	1.090	450	3.91	1.205	441	3.24	1.331	476	4.19	0.995	457	3.79	1.088
Russia	499	4.08	0.818	484	3.31	1.188	480	2.94	1.203	495	3.92	1.717	445	3.46	1.108
Sweden	481	3.70	0.867	475	3.89	1.060	471	3.14	1.025	477	3.84	0.971	476	3.72	0.984

Source: *What has been important in choosing the profession? (Give your opinion on a scale 1 [completely disagree] -5 [fully agree]).* (N = number of respondents, M = study means, SD = standard deviation). The "Journalism in Change" Project Survey, 2012.

7 Previous research by Nygren et al. (2010) indicated that 25 per cent of journalism students in Russia and 5 per cent of journalism students in Sweden want to work as journalists because they think it is profitable.

All of this highlights the differences between the Swedish and Central and Eastern European "model" for the role of income in the profession. Differences in economic conditions and situation in the job market serve as a background explanation for the data presented here. On the other hand, one could also ask to what extent tradition of a free press and journalistic autonomy might be used to elaborate on why journalists in some countries pay more attention to earning money rather than serving society? And last, but not least, following ideas advanced by one of the interviewees in Poland, all of this should be discussed in relation to journalists' perception of their income:

> "I think that people are encouraged, most of all, by thinking that it is a well-paid job, which is not true." [PR1, male, under 35 years old, radio, reporter, editor, chief of the new media department]

In Sweden, where media professionals did not pay a lot of attention to their salary when choosing the profession, only 3 per cent of Swedish respondents indicated being dissatisfied with the level of their income (responses for "*I have to get money from other jobs*") (Figure 3.8). The majority of respondents (78 per cent) indicated they could do well and provide extra things on the basis of income from journalistic activities. No significant differences between age groups and gender were observed in the Swedish case (Table 3.11 and Table 3.12). On the other hand, only 23 per cent of Polish and 17 per cent of Russian journalists agreed that they are fully satisfied with their earnings. More respondents in these countries said they could survive or provide for only necessary things when discussing the level of their salary. In addition, 14 per cent of Polish media professionals and 13 per cent of journalists in Russia indicated a need for getting money from other jobs. The situation was the worst in the group of the youngest and the oldest journalists surveyed (Table 3.11)[8]. In both countries more men than women were satisfied with their income (Table 3.12); this might be explained by the fact that more men than women are fully employed (see Table 3.9).

---

8 In the international project "Media Accountability and Transparency in Europe (MediaAcT)" 77 per cent of Polish journalists from the 20–30 age group indicated that an inadequate salary is a problem in the problem in a country (see Stępińska and Głowacki, 2014).

Figure 3.8: Evaluation of economic situation (in per cent)

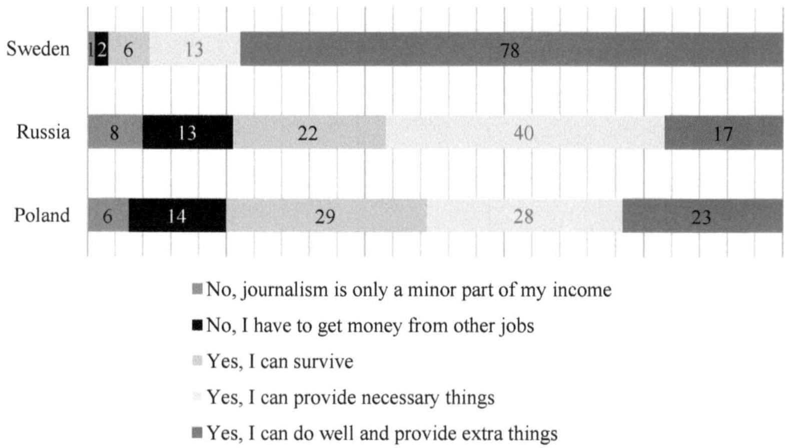

Source: *Is it possible to manage on your income from journalism?* The "Journalism in Change" Project Survey, 2012.

Table 3.11: Evaluation of economic situation of journalists with respect to age

Economic situation		*No, journalism is only a minor part of my income*		*No, I have to get money from other jobs*		*Yes, I can survive*		*Yes, I can provide necessary things*		*Yes, I can do well and provide extra things*	
		N	%	N	%	N	%	N	%	N	%
Poland	Under 35 (N 192 = 100 %)	18	9	31	16	60	31	48	25	35	19
	36–50 (N 176 = 100 %)	3	1	17	10	53	30	56	32	47	27
	Over 50 (N 83 = 100 %)	5	6	15	18	20	24	25	30	18	22
Russia	Under 35 (N 347 = 100 %)	36	11	50	14	69	20	133	38	59	17
	36–50 (N 115 = 100 %)	5	4	11	10	27	23	51	45	21	18
	Over 50 (N 36 = 100 %)	0	0	5	14	12	33	16	44	3	8
Sweden	Under 35 (N 125 = 100 %)	2	2	4	3	7	5	15	12	97	78
	36–50 (N 169 = 100 %)	0	0	3	2	9	5	18	11	139	82
	Over 50(N 175 = 100 %)	2	1	3	2	12	7	24	14	134	76

Source: *Is it possible to manage on your income from journalism?* The "Journalism in Change" Project Survey, 2012.

Table 3.12: Evaluation of economic situation of journalists with respect to gender

Economic situation		*No, journalism is only a minor part of my income*		*No, I have to get money from other jobs*		*Yes, I can survive*		*Yes, I can provide necessary things*		*Yes, I can do well and provide extra things*	
		N	%	N	%	N	%	N	%	N	%
Poland	Men (N 281 = 100 %)	12	4	35	12	69	25	89	32	76	27
	Women (N 203 = 100 %)	17	8	31	15	72	36	47	23	36	18
Russia	Men (N 188 = 100 %)	11	6	21	11	38	20	86	46	32	17
	Women (N 310 = 100 %)	30	10	45	15	70	22	114	37	51	16
Sweden	Men (N 233 = 100 %)	2	1	7	3	14	6	27	12	183	78
	Women (N 258 = 100 %)	2	1	3	1	16	6	36	14	201	78

Source: *Is it possible to manage on your income from journalism?* The "Journalism in Change" Project Survey, 2012.

Bearing all of this in mind, one could conclude that regular employment, full-time activities and adequate salary are provided to the majority of journalists in Sweden creates a secure work environment in which journalists do not need to think about basic financial needs, and can further engage in serving society and the public interest. However, journalists in Poland and Russia – who are not fully satisfied with their earnings – are getting involved in additional activity. They still attach a relatively high importance to the role of salary at work or analyzing reasons for entering the profession. An interesting question could be how do they confront their expectations when facing financial reality and benefit opportunities created by the employees?

3.5 Professional associations

Around the world different types of institutions and organizations based on a system of voluntary membership have been created to protect ethics at work as well as journalists' rights and interests in respective countries[9]. The overall aim of the professional association is usually to protect basic rights and journalistic freedoms. The goal is usually to support journalism identity and 'corporate solidarity', as well as professional culture based on principles of independence and media pluralism. In all three countries analyzed here different types of professional institutions have emerged, although differences in both membership and journalists' perceptions in Poland, Russia and Sweden are significant.

The Russian Union of Journalists (RUJ) is reported to be one of the largest unions of journalists in Europe with 85 regional organizations (*Russian Union of Journalists*, 2014). Its activities include participation in the process of law creation and promotion of journalism education. It has developed the Code of Professional Ethics of Russian Journalists adopted by the Congress of Journalists in Russia in 1994. The Swedish Union of Journalists (SJF) created in 1901 has the longest tradition here. It serves as a professional association and trade union, and gathers active journalists in the country. SJF stands for high standards of professional ethics and issues a membership press card, which is:

> "The only journalistic ID that guarantees that the holder has undertaken to comply with the ethical rules of the media industry." (*About the Swedish Union of Journalists*, 2014)

Finally, three different journalistic associations, namely the Association of Polish Journalists (SDP), Association of Journalists of the Republic of Poland (SDRP) and the Catholic Association of Journalists (KSD) continue to play the role in the Polish media landscape. The division of professional associations and rather 'hostile' attitude towards each other in Poland is rooted in the past[10]; both SDP and SDRP adopted their own codes of ethics and might be connected to right- and left-wing political orientation accordingly. The Catholic Association of Journalists was created in 1991 with the aim of promoting the ethics and role of Catholic media in the industry.

---

9 Professional associations operate also internationally; see for instance the International Federation of Journalists (ISF) and the European Federation of Journalists (EFJ).

10 The SDP was created in 1951 but was de-legalized by the Communist authorities in 1981 and further reactivated after the social and political transition of 1989. The SDRP was created in 1989 on the basis of the Association of the People's Republic of Poland.

The Russian Union of Journalists reports on its official website as having 70,000 members, while the number of active journalists in the Swedish Union of Journalists is estimated at the level of 17,000. In Poland the number of members of the SDP and KSD is 2,700 and 500 respectively[11]. Given the differences in the number of journalists in each country it is essential to have a closer look at the percentage of journalists who are the members of these kinds of organizations. By doing so, one will recognize the low popularity of membership in Poland and Russia. On the other hand, the extremely high importance of the SJF in Sweden will be observed.

According to Ghersetti and Weibull (2009) the SFJ almost doubled its membership in the period 1980–2000. The data from our project has shown that 89 representatives of the profession in this country belong to the professional association today (Figure 3.9). This trend is relevant in each age category and varies slightly from 86 per cent in the under 35 years old age group, to 90 per cent in the group of the oldest representatives of the profession (Table 3.13). On the contrary, both in Poland and Russia the majority of journalists (71 per cent and 77 per cent respectively) do not belong to professional union(s) or association(s) (Figure 3.9). In all the countries analyzed here (including Sweden) the highest percentage of respondents who are not members of professional association(s) is observed in the group of youngest journalists, which might raise several constraints on the future of journalistic associations. The 'gap' between the youngest and the oldest media professionals is the biggest in Russia. Looking at differences from the perspective of gender indicates that 30 per cent of men and 30 per cent of women are members of professional association(s) in Poland. In Russia where more than 60 per cent of journalists are women, only 20 per cent of female media professionals declare their membership at the RUJ. In Sweden, 87 per cent of female and 89 per cent of male journalists are members of the SFJ or other associations/unions (see Table 3.14).

11 Data on the SDRP has not been available.

Figure 3.9: Membership of trade unions/professional associations (in per cent)

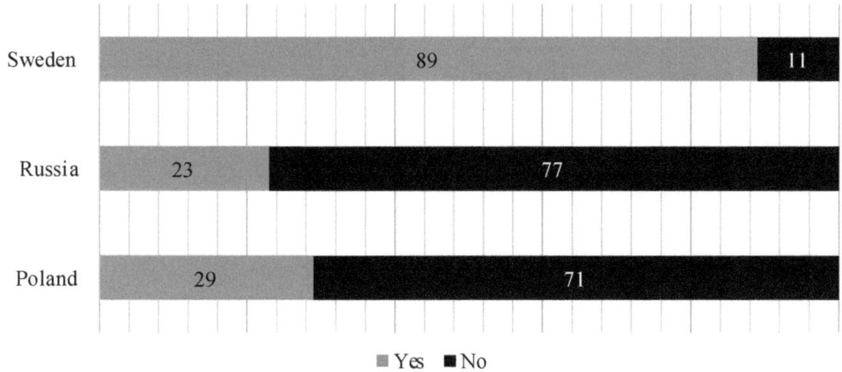

Source: *Are you a member of any professional organizations for journalists?* The "Journalism in Change" Project Survey, 2012.

Table 3.13: Membership of trade union / professional associations with respect to age

Membership of trade union / professional association		Yes		No	
		N	%	N	%
Poland	Under 35 (N 100 = 100 %)	18	18	82	82
	36–50 (N 100 = 100 %)	25	25	75	75
	Over 50 (N 55 = 100 %)	31	56	24	44
Russia	Under 35 (N 345 = 100 %)	52	15	293	85
	36–50 (n 114 = 100 %)	38	34	76	66
	Over 50 (N 36 = 100 %)	25	69	11	31
Sweden	Under 35 (N 78 = 100 %)	60	86	18	14
	36–50 (N 167 = 100 %)	149	89	18	11
	Over 50 (N 173 = 100 %)	156	90	17	10

Source: *Are you a member of any professional organisations for journalists?* The "Journalism in Change" Project Survey, 2012.

Table 3.14: Membership of trade union / professional associations with respect to gender

Membership of trade union / professional association		Yes		No	
		N	%	N	%
Poland	Men (N 159 = 100 %)	47	30	112	70
	Women (N 113 = 100 %)	34	30	79	70
Russia	Men (N 185 = 100 %)	53	29	132	71
	Women (N 310 = 100 %)	62	20	248	80
Sweden	Men (N 228 = 100 %)	203	89	25	11
	Women (N 257 = 100 %)	225	87	32	13

Source: *Are you a member of any professional organisations for journalists?* The "Journalism in Change" Project Survey, 2012.

There might be several explanations and discussions in relation to the study results, starting from tradition, long history, and benefits of being a member of a union of journalists in Sweden. Comments from one of the interviewees in Sweden are insightful:

> "It is important partly due to the fact that I get my press identification from there, which sometimes helps. Secondly, I get the newspaper "Journalisten" [The Journalist] which is pleasant to read. But above all, as a freelancer I am offered a lot of courses that are fun to attend and it's free." [SR16, female, 28 years old, freelancer and reporter newspaper]

While there is much that could be written about the reasons why Swedish journalists decide to join the SFJ, it is important to note that there are also some critical voices raised mostly by younger people and freelancers. One of the persons working for radio recalls his experience:

> "I am not in a trade union because I was so young when I joined the Dagens Nyheter and then there were those who were strongly opposed to my position against the union because of my age. It was about a personal relationship with the SJF. I think they have the wrong attitude when it comes to freelancers and those who are young. They live in in the print magazine world." [SR20, male, 41 years old, web editor, radio]

Another young person in Sweden agrees, by saying:

> "Well, just a few weeks ago, it was a friend who wrote on his Facebook page and asked if you should join the SJF, most respondents said "No: you won't get anything out of it." Around 15 people answered." [SR4, female, reporter newspaper, 28]

While the attitude of young people towards the functioning of the SJF might be changing, they are still not as critical as their Polish colleagues. Overall, journalists in Poland are very active when explaining why they do not belong to any type of journalism association. Among the main reasons one might mention is distrust, the lack of benefits as well as core activities and impact on journalism (Dobek-Ostrowska et al., 2013). The existence of different associations with clear political ties in Poland might be treated as a factor responsible for negative assumptions, which – in some cases – include questions about the legitimacy of the existence of these associations:

> "Trade Unions don't concern me, and the SDP – as far as I know – is strongly politically involved which doesn't suit me at all, in principle." [PR19, male, 35 years old, public service TV]

> "Trade unions and journalism associations make no sense to me; they're of no use to young people, they just don't represent anything. And nowadays the only thing people associate trade unions with is their political involvement, and it's just pointless. They never achieve anything in private companies; they have no right to exist really." [PR15, male, 35 years old, private radio]

> "I am interested in my profession but I don't feel the need to belong to any organizations. Somehow it is hard for me to believe that they can help me when something bad happens to me." [PR8, female, 35–50 years old, public service TV]

Bearing in mind all the concerns mentioned above and our study limitations it would be further interesting to dig seriously into reasons why people decide to join these associations and further pose a question on the role of political ties and sympathy when joining a union. Does the fact that older journalists in Poland are more likely to belong to professional associations than younger journalists make any difference in the reporting of older versus younger journalists? Also it could be worth exploring to what extent people who are the members have done this for reasons related to job security rather than from improving service to society? As argued by one of the media professionals in Poland:

> "They [journalists – MG] are afraid they can lose their jobs and they hope that if they belong to such a union it will be more difficult for an employer to fire them. It is a notion like that which is deeply rooted in people's consciousness but it is not true." [PR1, male, under 35 years old, radio, reporter, editor, chief of the new media department]

Finally, although our research data does not allow to discuss this, it might be of high importance to measure the dynamics and changing membership to fully understand the tendencies; it could be helpful to have a closer look at the perception of professional association(s) from the perspective of existing members; one of the interviewees in Poland claims:

> "Formally I have belonged to the SDP for years and as a matter of fact through laziness I haven't discontinued my membership. But I should." [PR3, male, 55 years old, national weekly]

Although Russian journalists are not that active when revealing and commenting on their attitude towards professional associations, one might expect that the reasons related to independence and impact might be similar to the case of Poland; as argued by Lauk with a reference to Central and East European professional associations:

> "The inability to establish an organization capable of protecting the independence and autonomy of the profession and ensuring internal control over the quality of journalistic performance signals the absence of a shared professional ideology and interests." (2009: 73)

In Russia, where about 75 per cent of members joined a professional association before 1992 (Pasti et al., 2012), younger people might simply not see any need of being affiliated with any type of institution or they might be dissatisfied with the operations of the RUJ.

3.6 Conclusions

The overall aim of this chapter has been to recognize and analyze selected dimensions for comparative studies on journalism. The opening discussion on "Who is a journalist today?" highlighted the complexity of approaches and different factors that might be helpful to understand journalists' features in the era of social change as well as new media and technologies. Our goal has been to propose a framework for the analysis, which is treated here as a background for detailed discussions and studies of journalism in Russia, Poland and Sweden. By applying selected demographic variables – gender and age groups – we have explored three areas of the "Journalism in Change" study, paying particular attention to education and training, employment and salary, and journalistic perception of trade unions and professional associations. All of this helped us to observe characteristic features of journalism in each country. In addition, it served as a reference point for mapping similarities and differences between them.

Firstly, then, countries are different with respect to gender and age groups proportions. Among the journalists surveyed in Poland there are more men than women; the biggest group of journalists (42 per cent) were under 35 years old. More women than men surveyed have indicated having a degree in higher education; the highest percentage of respondents with the degree belonged to the generation of the youngest people. Poland is an example of a country with a relatively high percentage of journalists without a degree in journalism (43 per cent). At the same time, there is a relatively large group of young people having a degree in journalism (more than 50 per cent when taking into account both academic and non-academic level). The data gathered from the survey might be explained by the quality and tradition of professional training in the country, where – despite the growing number of schools and universities offering courses on journalism education – some journalists still think that more specialized knowledge about the world is more important. Given the fact that in several cases more attention has been put to in-house training in media organizations rather than a degree in journalism, the broader picture in a country might also be explained by expectations of future employees who might want to put more emphasis on practical skills and learning journalism on the basis of daily functioning of a given media enterprise. In addition to this, as data related to employment indicates, the situation of the youngest representatives of the profession is the most difficult: 49 per cent of them have regular employment in media companies and 67 per cent get full-time jobs. Journalists in Poland see economic pressures as a factor that might have a negative impact on the quality of journalism in a country. Salary plays an important role across all the journalists' generations; due to their job market positions the youngest people are the most dissatisfied with their salary and some of them have second jobs. Finally, in Poland there is distrust towards membership in trade unions and professional associations. Less than 30 per cent of Polish journalists indicate belonging to one of the three associations; more than 50 per cent in the age group over 50 years old belong to journalistic association(s). Younger people do not want to affiliate themselves with any types of organizations; the existing structures have been widely criticized for being divided and politically oriented.

One of the most visible features of journalism in Russia is the dominance of female professionals. The majority of journalists surveyed were from the under 35 years old age group, which might play a significant role when understanding Russian journalists' attitude towards new technologies, the existing political regime and communist past of the country. Overall, there have been many negative assumptions when evaluating the quality of journalism of the youngest generation of journalists in Russia; they have been criticized for being unskilled, demotivated and politically-oriented. At the same time due to prestige and/or job market competition, Russian journalists are well educated – 96 per cent of them have high education and, this needs to be underlined – 75 per cent of respondents have obtained a degree in journalism. The situation in the job market in Russia could be regarded as rather stable, since the majority of journalists have regular employment at a media company and 68 per cent of them work full-time. Similar to the Polish case, Russian journalists recognize the growing role of the economy when evaluating current developments of media and students of journalism choose the profession either from the perspective of professional development or financial benefits. Salary plays an important role when choosing the profession; more women than men and almost 25 per cent of the youngest people are dissatisfied with their income. There is a relatively high proportion of people who declare that journalistic activity generates just a small proportion of their income; approximately 20 per cent of journalists in Russia might have the other jobs to meet the standards of living. Generally, Russian journalists are not interested in membership of professional associations (this does not apply to the over 50 years old age group, in which 69 per cent declared a membership of a trade union/journalistic association).

Finally, the Swedish case is different from the Polish and Russian one in several ways; one of them is the fact that the biggest group of journalists surveyed are from the age group of more than 50 years old. As opposed to the Polish and Russian cases there has not been a clear generational shift in relation to social and political transformations. Swedish journalists are more experienced and the largest age group entered the profession during the expansion in the 1970s and1980s. There are more female journalists in the country, and the gender differences have been the biggest among the youngest journalists. This has been explained by the expansion of journalism education, which plays a significant role when defining journalism in the country. Swedish journalists are well educated (90 per cent with higher education), flexible, and the majority of them have education in journalism either from university or school of journalism. Generally, journalists in Sweden are mostly employed on the basis of regular contract and job security is more important for them than salary. In addition to this, journalists are very satisfied with their income. Since more than 90 per cent of people working as journalists full-time and only 2 per cent need to take an additional job to survive financially, one can conclude that the majority of media professionals in Sweden have the comfort of investing all their professional time in journalism activities. This might include being actively involved in operations of journalistic association (89 per cent of media professionals indicating membership of the SJF), which creates a plethora of benefits and is still extremely popular regardless of gender or age.

There are different approaches, which might be used to explain all the similarities or differences between all the cases analyzed. Without a doubt, Poland, Russia and Sweden represent different journalism cultures and several contextual factors (press freedom, tradition, media policy, media-political relations, economy, the development of professional standards) might be helpful in understanding national characteristics and features. The Swedish model of Swedish journalism is treated exclusively due to a long history of autonomy, responsibility, identity, as well as a more stable economic situation than in Poland and Russia, and a strong need for being a part of journalistic community. Post-communist Poland and Russia are similar in many ways, although some of the differences between them are salient. These have a direct and profound implication on different attitudes towards journalism education as well as employment of young people representative of media organizations today. In all the cases described above, looking to the past cannot be made without reference to current changes in the profession – the growing role of market competition and impact of the economy, the changing roles of journalism, and changing relations with the public in the new media ecology regime. Journalism is nowadays changing around the world; this creates the situation in which traditional "borders" between Western and Eastern approaches to the profession might have been "blurred".

The aim in this collection is to observe to what extent traditional features and differences with regard to structures have contributed to today's understanding of professional norms and values, growing market competition, changing journalism roles and the rise of network practices. All the preliminary remarks and discussions will therefore be used to understand and explain more detailed analyses in the following chapters of this collection.

4. Changing working conditions

Jöran Hök

4.1 An organizational approach to journalism studies

Studies of working conditions are vital while researching preconditions for journalism and changes within the profession. Studies of this kind are important as they identify the fundamental daily conditions under which news is produced.

Studying working conditions is part of an organizational approach to journalism. The concept is often related to collective relations as employment conditions, arrangements within a profession that are regulated in a contract between employer and employees or in agreements between unions and companies. On the individual level an organizational approach focuses on:

> "(...) mutual expectations between the individual and the organization, ranging from hard facts, such as salary, to softer criteria, such as job satisfaction" (Altmeppen, 2008:53).

Organizational studies of working conditions focus on the number of staff, necessary skills, working-hours, numbers of assignments during a working period (production pace) and job-security. But the concept should also be related to a working culture in a broader sense, including such vital professional conceptions as space for creativity and other aspects of perceived quality in the work place.

This chapter compares five aspects of changes of working conditions as perceived by the individual journalist in the three states:

(1) The first research question concerns how journalists perceive changes relating to the size of the work force and its implications for the quality of work satisfaction and work output.
(2) The second research question concerns perceived changes relating to the amount of skills needed in daily work.
(3) The third research question concerns perceived changes relating to the produced quantity within a working day and changes in time limits for delivery of articles or stories; i.e. workload and time limits of production.
(4) The fourth research question concerns changes related to perceived possibilities to influence workplace conditions.
(5) The final research question concerns work place mobility; i.e. how professionals view the willingness to stay on within the profession or moving on to another job.

4.1.1 Changes in size of workforce

Changes in the size of journalistic workplaces in each country give valuable information of whether a profession is expanding or contracting. Usually expanding professions increase opportunities for higher standards in working conditions and higher wages, but there is no evident correlation between higher numbers of employed and increased quality in work.

The financial crisis of 2008–2009 hit the media labor market in all three states. According to research in Russia after the economic world recession of 2008 a third of the journalists in Russia lost their jobs (Pasti, Chernych and Svitch, 2012:269). Polish journalists have passed a "phase of erosion" (Dobek-Ostrowska et al., 2013: 37.) with staff cuts of nearly 50 per cent within a ten year period and around 25 per cent in the last three years (Dobek-Ostrowska et al., 2013: 37) which is at a faster pace than in the U.S., where the cuts stayed at around 30 per cent from 2000 to 2012 due to the yearly statistic report from the Pew Institute (2013). The Swedish situation resembles the Polish decline although at a slower pace. In the decade 2004–2014, newspapers reduced their staff by about 25 per cent, and commercial TV and radio probably more (Nygren and Althén 2014: 26). Staff reductions have been most severe during 2013 and 2014, mostly in printed media and among these in daily newspapers and are explained by shrinking advertising revenues and shrinking numbers of subscribers.

The downsizing is visible in our survey, but the trend in Russia is different in comparison to Poland and Sweden. Nearly 70 per cent of the Swedish respondents say that editorial staff has been reduced in the last five years, and in Poland the same figure is nearly 60 per cent. In Russia around a third of the respondents say that they have experienced shrinking staff, but also 25 per cent see an increasing staff. However, the figures from Russia are unreliable, as almost a quarter of all respondents answer "do not know" to the question (see Table 4.1, 4.2 and 4.2).

Table 4.1: Changes in the size of editorial staff in total

Country	Shrunk (1–2)		The same (3)		Grown (4–5)		Don't know /no answer	
	N	%	N	%	N	%	N	%
Poland (n 482 = 100%)	281	58	97	20	72	15	32	7
Russia (n 465 = 100%	151	32	95	20	114	25	105	23
Sweden (n 410 = 100%)	281	69	35	9	54	13	39	9

Source: *During the last five years, has the size of your editorial staff grown or shrunk? (Answer on a scale 1–5 shrank – grown).* The "Journalism in Change" Project Survey, 2012.

Table 4.2: Changes in the size of editorial staff at newspapers

Country	Shrunk (1–2)		The same (3)		Grown (4–5)		Don't know / no answer	
	N	%	N	%	N	%	N	%
Poland (n 125 = 100%)	103	82	12	10	4	3	6	5
Russia (n 151 = 100%)	53	33	28	20	34	22	39	25
Sweden (n 200 = 100%)	175	88	11	6	5	2	8	4

Source: *During the last five years, has the size of your editorial staff grown or shrunk? (Answer on a scale 1–5 shrank – grown).* The "Journalism in Change" Project Survey, 2012.

Table 4.3: Changes in the size of editorial staff in the public service media

Country	Shrunk (1–2)		The same (3)		Grown (4–5)		Don't know / no answer	
	N	%	N	%	N	%	N	%
Poland (n 104 = 100%)	60	58	19	18	21	20	5	4
Russia (n 50 = 100%)	12	24	16	32	16	32	6	12
Sweden (n 72 = 100%)	44	61	6	8	8	11	14	19

Source: *During the last five years, has the size of your editorial staff grown or shrunk? (Answer on a scale 1–5 shrank –grown).* The "Journalism in Change" Project Survey, 2012.

If we compare different workplaces, we get a clearer pattern of differences of perceived changes of staff. The largest group of all respondents in the survey belongs to the category of printed media and here the perceived reduction is prominent in all countries. In Poland 103 out of 125 (82 per cent) said that they have experienced shrinking staff in their newsroom in recent years. In public service, radio and television, 60 out of 104 (58 per cent) said the same. Experiences of increasing staff were less: 18 per cent or 21 persons. In Sweden as many as 175 out of 200 (88 per cent) working in printed media say that they have experienced shrinking staffing in their newsrooms. In public service media 44 out of 72 (61 per cent) expressed the same view. Experiences of increasing staff were fewer: only 5 persons or 4 per cent in printed media and 21 persons (30 per cent) in public service media worked in companies with expanding news staffing. In Russia, 53 out of 160 in printed newspapers (33 per cent) have seen shrinking newsrooms, while 34 persons (22 per cent) had the opposite experience. But, as mentioned earlier, many could not answer the questions because as many as 29 persons (25 per cent) answered, "do not know" to this question.

The answers give an indication of differences within and between countries. The over-all impression is that we can see an uneven development, where printed media loses out, even if Russia to some extent experienced a smaller reduction after the recession of 2008.

Two important connected factors seem relevant in explaining the situation. Firstly, it can be related to a general crisis of the business model. Advertisers have left paper media for other media platforms and circulation of printed media is slowly decreasing, due to shifting habits among advertising agents and companies and simultaneously among media consumers (Levy and Nielsen, 2010:5). This trend is striking in all three states. It can be related to the rapid changes in media habits, where younger generations do not subscribe to newspapers in the way older generations did, and instead prefer on-line news and feature journalism. Secondly, it can be related to the fact that companies generally have been reluctant and half-hearted to follow advertisers and consumers in a reorientation towards on-line media. When this reorientation finally has taken place it has been slow and the revenues have not been sufficient, which has had the implication that companies have been forced to decrease costs by reducing staff.

4.1.2 Changes in size of workforce and the implications for the quality of work

The development of labor markets gives us certain clues concerning changes in the conditions for daily work. The general trend in North America and Europe has been a declining work force. In many states there is also a tendency towards so called "flexible" working conditions where the companies rely more on temporary workers, freelance workers and/or production companies (Deuze, 2007: 19; Davies, 2009:63; Fenton, 2010:15; Peters and Broersma, 2012:6).

One of the Swedish respondents can illustrate the sentiment of many who have experienced setbacks in the profession:

> "It will continue as now: the major media companies are cutting down. It will be an insecure job market for journalists. (...) These tougher conditions will be reflected in the journalistic product."[SR 14, female, 37 years old, magazine reporter]

Another reporter talks about the effects of cutbacks in terms of change of resources:

> "We do not get more resources, which means that when you yourself do the entire segment this takes far longer. One needs to shoot, do research, write web text and post it on the Web. And if you downsize the employees then the work will be of lower quality and

when you produce more what you see on the screen will be of lower quality." [SR5, male, 30 years old, video reporter]

And finally, a third of the Swedish respondents describe the changes in terms of implications for investigative journalism:

"I want investigative journalism to remain, not that it should be removed. The reason for my concern is that you can see that some newspapers may cut down. Other cutbacks include the quality of journalism. Public service in Sweden is under investigation and I am concerned about what it lands in terms of resources and limiting basis" [SR 2, female, 35 years old, television reporter]

A cross-tabulation among respondents to the survey who have experienced cutbacks and made a statement on quality shows if these journalists see a connection between cuts in staff and quality of work.

Table 4.4: Connection between views on changes in quality in journalism among respondents who have experienced decreasing or increasing staff in the newsroom

Country		In total		Quality decreased (1–2)		The same (3)		Quality increased (4–5)	
		N	%	N	%	N	%	N	%
Poland (n 441 = 100%)	Fewer staff	257	58	191	74	43	16	23	9
	More staff	69	16	41	59	21	30	7	10
Russia (n 444 = 100%)	Fewer staff	146	33	70	48	57	39	19	13
	More staff	109	25	54	50	26	24	29	27
Sweden (n 350 = 100%)	Fewer staff	249	71	105	43	84	34	60	24
	More staff	46	13	21	46	13	28	12	27

Source: *Do you think the quality of journalism in your country has increased or decreased in the last 5–10 years?* (Answer on a scale 1–5 decreased –increased). *During the last five years, has the size of your editorial staff grown or shrunk?* (Answer on a scale 1–5 shrank –grown). The "Journalism in Change" Project Survey, 2012.

Table 4.4 shows that most of the respondents say that they have perceived a decreased quality of journalism. This holds true even if staffing has increased in the respondent's newsroom. It is notable that this also is the case in Russia, where the respondents have experienced a relatively high increase in newsroom staffing compared to the two other states. Due to these figures the respondents seem to suggest that quality in journalism is decreasing in all kinds of newsrooms.

However it is possible to obtain a better quality even if you reduce staff. This can be the case when you use more freelance material or buy material from production companies, a situation which is already in place. This can explain why many respondents in Russia and Sweden say that quality has increased although staffs have been reduced in their workplaces.

To conclude: there has been a significant reduction of the workforce in the three countries after the recession in 2008–2009. The only difference is that the survey responses indicate a slight increase in the newsrooms of broadcasting in Russia. A reduction in staff in newsrooms need not in itself lead to a reduced quality of the output, but the surveys and interviews suggest that the journalists in the three countries say that the quality decreases and that this is related to both downsizing of the workforce and increasing demands on production for different channels. One can speculate on what impact this may have on journalism as a democratic watchdog, especially concerning the consequences for investigative journalism, which is labor intensive. Some interviews provide justification for the hypothesis that journalists perceive journalism in this regard is weaker today if you compare the situation now with ten years ago.

4.2 Changes towards multiskilling

Journalists are to a large extent working for many platforms; printed media, web, television and radio – at least two of these. There is a general trend to prefer knowledge of "multi-reporting" when hiring new personnel. This is a part of media company strategies to expand into other platforms when advertisers and readers leave traditional newspapers and magazines. To value the extent of this change, we can use radio as an example. For a radio reporter, writing has become more central, while the traditional occupation, quite different in style and format, now is put in second place. The reporter has to plan for and deliver it in two formats. One Swedish radio reporter describes the new environment:

> "Everyone needs to circulate, we do radio, take pictures and post on the web. Often, the web comes first, it is important to remember that and to know what fits there (…) These change have occurred in the last four to six years, before that it was always the radio first. When the Web came into being, it was perceived as a burden that took resources away from our real jobs instead of being a complement." [SR 2, female, 35 years old, television reporter]

A middle-aged Russian reporter working at a news agency has the opposite viewpoint:

> "Almost all of my colleagues, me included, went through the system of multimedia education. We were not only taught how to write, but also to film and video edit. Of course, multitasking is crucial. First, in any media a situation of "replacing your colleague" occasionally arises because of some force majeure. Second, multitasking helps to compete with other media and media enthusiasts. It's good to be multifunctional when it was the journalist's own choice, if only they manage to combine. Another thing is when life makes you work in different media and fulfil a lot of duties. For me personally, multitasking is good, for I cannot put myself into a rigid framework." [RR6, male, 44 years old, reporter working for a news agency]

A young Russian TV producer says:

> "Our TV Company is famous for "doing everything", most employees don't have a specific position… a person can be a reporter, an editor, there's a busy rotation… All employees of the channel are in some sense multifunctional." [RR20, female, 25 years old, television reporter]

A Swedish TV reporter in the youngest age cluster describes the implications of multiskilling for reporters at his company:

> "The advantage is that if everyone is multi skilled, you can plan the schedules in any way, you're not stuck in roles, we can help each other with different things. I can shoot while another reporter is doing something else. Previously, you had to opt out of something, but now a reporter can also shoot, edit and interview, then we can get more out of it. The negative is that we do not get more resources, which means that this takes longer time". [SR5, male, 30 years old, video reporter]

In the survey a majority of the respondents answered: "agree", to the statement: "I feel like a multiskilled reporter". The respondents in general did not agree with the statement that multiskilling decreases journalistic quality. Most respondents were in favor of multiskilling enhancing creativity and "power" to individual journalists. Both these categories can be seen as vital parts of professional self-esteem. They can also be seen as an indication of the similarities and differences in the development of journalistic cultures in the three countries. The attitudes toward changes in the direction of multiskilling differ when comparing some answers from the three countries, which is shown in table 4.5.

Table 4.5: Attitudes towards multiskilling in the three countries (mean on a scale 1–5 where 5 is "fully agree"))

Country	*I feel like a multiskilled reporter*	*Multiskilling gives more space for creativity*	*Multiskilling will decrease the quality of journalism*	*Multiskilling gives more power to the individual journalist*
Poland (N 450–482)*	3.9	3.8	2.8	3.7
Russia (N 473–494)	3.2	3.6	3.0	2.6
Sweden (N 456–476)	2.8	3.0	3.1	3.9

*Note that N is different in each country, as respondents did not answer all statements.
Source: *"Multiskilling" or "multi-reporter" are the words often used to describe future journalism. It can mean both a journalist working in different channels and a journalist working with different parts of the process (photo-writing-editing). What is your opinion on the following statements*? The "Journalism in Change" Project Survey, 2012.

The presented data (Table 4.5 and Table 4.6) shows that Polish and Russian respondents are more positive towards multiskilling. Two younger Polish voices, a radio reporter and a magazine reporter, can be seen as representatives of this generation:

> "There are only advantages. By multitasking we prevent ourselves from rusting. You can have endless possibilities. The more tasks we coordinate, the more fulfilled we feel. We're not cogs in a machine." [PR 13, female, freelance radio reporter]

> "Very often it is said that if somebody is good at everything, then in fact, he/she is good at nothing. I think it is great to be good at everything at least partly. (…) I think that multiskilling is an advantage. Personally, I can't think of any disadvantage." [PR 20, male, 35 years old, magazine reporter]

On the contrary, Swedish respondents in general seem to be more negative toward this development, which can be a consequence of the more elderly profile of the work force in Sweden. Differences between generations are visible if we look at the categories (1) "multiskilling means more space for creativity" and (2) "multiskilling gives more power to the individual journalist." Table 4.6 shows that younger journalists to a higher degree answered agree and strong agree with these statements than older journalists.

Table 4.6: Positive attitudes towards some statements on multiskilling among different generations (in per cent)

Country	Under 35 years old		36–50 years		Over 50 years old	
	Creativity	More power	Creativity	More power	Creativity	More power
Poland N243/ 269	57	58	56	61	47	70
Russia N 226/250	51	56	37	48	28	31
Sweden N 234/142	56	35	51	33	48	31

Source: *The respondents answered 4 or 5 on a scale 1–5 to the statements that multiskilling "means more space for creativity" and "gives more power to the individual journalist".(Give your answer).* The "Journalism in Change" Project Survey, 2012.

In Poland there is a visible majority who connect multi skills with creativity. The same pattern was visible among Swedish respondents, where 74 out of 116 (56 per cent) of the youngest group agreed with the statement that multi-skilling enhances creativity, while slightly fewer did agree in the elderly groups. The Swedish figures reflect the more sceptical view on multiskilling, which was referred to earlier in this chapter.

A comment from a younger Swedish freelance journalist, mostly working for a daily newspaper, can describe this state of art:

> "I think many see the younger generation as a threat in the industry, given that we are often multi journalists. We do everything, while many older journalists might have written for print throughout their career, and certainly aren't going to learn how to write on the web or doing web TV. It is the direction everything is going, some might see it as an advantage to have an input of people who know other things too. But I think many see it as a threat because a lot has happened in how we present journalism. The older generation isn't as updated in the same way." [SR 16, female, 27 years old, freelancer]

Finally, among the Russian respondents the tendency is not that strong, though the number of younger respondents is much more prominent than in the other countries. However, Russian respondents think that multiskilling gives individuals more power in the newsroom. This can be seen as a sign of hoping that professionalization will promote autonomous work within an authoritarian media landscape.

Although multiskilling is an increasing phenomenon, the general views on the advantages or disadvantages of multiskilling are polarized among the respondents in the interviews in each country. In Poland, a third of the interviewed are negative towards the transformation to multiskilled journalism. A third of the interviewed, mostly the younger, are positive. But there are also different opinions among the group under 35 years. A younger Polish radio reporter says:

> "If somebody is to be good at everything, it means they are in fact good at nothing, I mean, nowadays, media concerns put great emphasis on the so-called 'combines' or people who can take photos, can make a video and can also prepare some good material, meaning an article for a newspaper, the radio, or the Internet. It of course brings with it what it brings, i.e. content-related, factual errors, because there is not enough time to prepare well for the article, it brings with it, apart from the factual errors, obviously, some stylistic errors, the articles are perfunctory, usually constructed on the basis of press information. (…) The advantage of journalists being kind of multiskilled, well, might of course be that their superiors consider such guys to be money making machines." [PR1, male, editor, radio]

A more ambivalent view is expressed by PR 15, also a younger radio reporter:

> "The advantage of multitasking is that it translates into a higher citation rate and you can write more. The radio is a limited form of transmission; the programs don't last the whole day. They're broadcast the entire day but last five minutes instead of thirty, and this is what makes working in radio difficult. Therefore, multitasking gives us the opportunity to produce in a more detailed way. The disadvantage is that it takes up a lot of time, and if we don't prioritize, we won't have time to do other things, like chasing the news, which should be the most important. And of course, you can get fed up when you leave work every day dead tired, it kind of takes away the joy of being a journalist and can cause problems, so it's worth taking a step back from multitasking and relax a little every now and then." [PR 15, male, reporter, radio]

Among the persons interviewed in Russia the impression is that multiskilling journalism is a rather new phenomena. A young Russian television editor says:

> "We don't have any absolutely converged media yet, so there's no duty to be absolutely multifunctional." [RR2, female, 24 years old, editor television]

Most of the respondents agree that this changing environment is impending. A young television reporter says:

> "I believe that a journalist has to be multifunctional. For example, the Faculty of Journalism today teaches journalists both to shoot and edit, and it's great. In case something happens to the camera operator, the journalist will be able to replace him/her." [RR11, male, 24 years old, reporter, television]

A few of the reporters are not so convinced. One of them says:

> "I don't understand the idea of multi-functionality at all, I absolutely don't understand the popular saying "a journalist has to be a jack of all trades". For example, to prepare an expert text for someone who is out of subject is absolute nonsense. Journalism must be specialized." [RR 1, male, 35 years old freelancer]

In Sweden, a web editor, points at some problems for specialization:

> "One can become proficient at understanding how content is communicated in various formats, which can be useful. But when the publishers are larger, like Swedish Television (SVT), I think it's the wrong direction to go, everyone does everything and knows everything. In the past they had specialists in all fields. They worked in teams. But with digitizing, there are tools so that anyone can do almost anything. Then everyone learns it for some strange reason I do not understand, but it seems to be about a belief in efficiency. But I do not believe in it, I believe that there should be specialists. I think you would get both better and more media if they were all specialists. That one retained the old way where a sound engineer really is a sound engineer and that you support each other. Some find that this line that we have today is the smartest, but I do not think so. The best final format is the most important. I think the final delivery will be better if we are all specialists." [SR 20, male, 40 years old, web editor, radio]

A Swedish web editor problematizes the effects of digitalization on the working process:

> "Overall, digitization is a misconception for many. They believe that it provides an improvement, but there is actually nothing that says it makes something better. (…) It makes it cheaper to deliver content. It has allowed the publishing houses to earn even more money. But it has degraded quality in almost all areas it has entered (…) Digitalization itself is not good. It is linked to this idea that everyone should do everything just because there are digital tools that allow you to do it. But that does not account for an improvement in itself." [SR 20, male, 40 years old, web editor, radio]

Concluding, the rapid development of multiskilled journalism is one of the most pervasive changes to the working conditions of journalism in recent decades. This convergence in the workplace places greater technical demands on the individual reporter. It also requires greater expertise of how to work within different formats, such as dissimilar narrative techniques. We are in the midst of the process of change to multimedia journalism in all countries and development has been unequal. Opinions on the change are also different, as shown by some of the cited interview responses. However, the survey provides justification for the hypothesis that the development of a multi-media journalism is considered irreversible. Of those surveyed, the majority answered that multi skills give more room for creativity. To this aspect the survey shows a uniformity of opinion between generations. However, the views are more divided regarding if multiskilling gives more power to the individual journalist. The respondents are more likely to affirm that statement in Russia and Poland, while in Sweden the respondents express greater scepticism. This disparity between the countries can be an expression that multiskilling is seen as a form of up-skilling in Poland and

Russia, while Swedish journalists to a larger extent see this development as an expression of rationalization mainly caused by financial difficulties.

4.3 Changes in workload and tempo

Modern information technology has changed the way in which journalists retrieve information and the way in which they are in contact with sources and audiences. Telephones and "shoe tearing journalism" have to a large extent been replaced by staying at the workplace using computers for communication. Reporters produce more articles and programs per person compared with ten years ago. They are often urged to publish rapidly using web publishing to get ahead of the competitors.

A middle-aged Polish radio reporter says:

> "Here you have seven days and the closer to the end of the seven day period the quicker you have to work." [PR4, male, radio]

Another reporter from the same age-group discuss the causes of increasing speed in producing news material:

> "Everything has speeded up. In the past 20 years everything's been changing so fast. Media try to be at the scene immediately. There is such a technological possibility..." [PR 12, male, reporter newspaper]

In the survey the respondents were asked to evaluate changes within the profession connected to daily work. Table 4.7 shows tendencies in general in the three countries of these changes:

Table 4.7: Evaluated changes in different areas of daily work the last 5–10 years (mean on a scale 1–5 where 5 is "much more")

Country	Writing/ producing	Editing/ layouting	Amount of produced material	Evaluation and discussion
Poland (n 369–441)*	3.53	3.99	3.34	3.06
Russia (n 350–409)*	3.50	3.93	3.33	3.39
Sweden (n 326–380)*	3.50	3.45	3.96	2.75

*Note: only those with 5–10 years of experience could answer the question

Source: *Over the last 5–10 years daily journalistic work has changed in many ways. How do you evaluate these changes in different areas, according to your own experience?* The "Journalism in Change" Project Survey, 2012.

While evaluating changes in the amount of "writing and producing" texts or programs the tendency is the same in the three countries: it has increased. The level of working with 'editing and layout' has also increased in all three countries but is more prominent in Poland and Russia. When the "amount of production" is estimated, Swedish journalists in general say that it has increased, as 71 per cent of Swedish respondents say that there is more of this, while only 6 per cent answer "less". Swedish journalists also respond differently to the important question of evaluation and discussion, *as well as the implementation of work*. Forty-two per cent of Swedish respondents say that there is less evaluation and discussion today when looking back over the last 5–10 years.

A younger Polish newspaper reporter elaborates on the situation:

> "In the past you sat down, gathered your material slowly and you wrote a longer article and a journalist was seen more as a writer. Now it is all much faster and for example a press article needs to have its online version immediately. Because of that, we need to summarize and re-formulate press articles so that they can be used as Internet news, so we need to divide our time, we need to work on two things simultaneously and it also takes up the time needed for writing a longer article." [PR 2 female, 30 years old, reporter, newspaper]

A young Swedish TV reporter TV reporter says:

> "In the three years I've worked here, our channel has changed a lot, it's much more stressful today. We've had a change in technology. When I arrived you had to bring the tapes and transfer them to the computer yourself, then you got a natural break in your workday. But now you put a card into your computer, the material is instantly stored digitally and then you can edit right away. Everything goes faster but it also raises the demands on yourself." [SR5, male, 30 years old, video reporter]

A young Polish radio reporter says:

> "You can get fed up when you leave work every day dead tired; it kind of takes away the joy of being a journalist…" [PR 15, male, reporter, radio]

If we look at one of the changing factors in Table 4.8 (changes to the working tempo), the survey provides insights of how individual journalists judge effects of increasing working tempo for independent journalism. The results indicate that increasing working tempo is seen as a threat to journalistic independence in all three countries, but most significantly in Poland and Sweden.

Table 4.8: Views on threats to independence in work by increased tempo in the workplace (in per cent)

Country	No threat (1)	2	3	4	Large threat (5)
Poland (n 481 = 100%)	2	6	20	47	26
Russia (n 494= 100%)	8	9	28	36	19
Sweden (n 485= 100%)	2	3	16	44	35

Source: *Journalistic independence can be influenced by many factors. How do you value the following factor: Increased working tempo* (1: no threat– 5: large threat). The "Journalism in Change" Project Survey, 2012.

When working tempo is increased journalists are producing more material, more often and within shorter time limits. This is considered as problematic relating to the possibilities to verification as there is less time for fact-checking, at least in a shorter time perspective. A young Russian TV-producer says:

> "The disadvantage of this profession is that material has to be presented promptly; we often don't have time for in-depth analysis of the material. We don't even have a clear-cut line between subjects… " [RR20, female, 25 years old, reporter, television]

A Polish radio journalist says:

> The content of the transmission has become more superficial. It's because of that herd instinct phenomenon that happens every now and then. This means that, whenever anything new happens in a particular media, everyone turns to it and starts dealing with the same issue. So the quick access to a particular place has advantages and disadvantages. It's not good when there's no variety. In my opinion, it's wonderful that people want to know everything, but it's important that they know more, not only what happened, but how it happened and the consequences that there will be. They need to know what this will mean for the future." [PR 12, male, reporter, radio]

An older Swedish newspaper reporter says:

> "The problem now is to find enough time. With compressed working hours, there is no room in the organization for your own initiative. One must perform such things during overtime or time off, just because you think it is fun. You have to take a night without getting paid. The magazine accepts ideas and suggestions, but the working hours… When I previously edited I did not always get paid if I worked outside working hours. They were unwilling to pay overtime. That leaves your off hours." [SR 1, male, 59 years old, newspaper reporter]

If we pick out three important working conditions from Table 4.7 and Table 4.8 we can construct the following picture of perceived changes for today's journalist in Poland, Russia, and Sweden respectively:

Table 4.9: Comparing changes in working conditions in three countries

	Poland	Russia	Sweden
Working tempo	Higher	Higher	Much higher
Workload: amount produced	Higher	Higher	Much higher
Space for discussion and evaluation	The same	Higher	Lower

Source: Author

The explanation for the different outcome in Sweden can be that Swedish professionals experience a more rapid decline in combination with a weakening of the possibilities to influence working conditions, which is the next aspect of changing working conditions to be discussed.

Concluding, requested to compare the situation today and ten years ago, journalists in all three countries experience that fewer people produce more in less time. This can be seen as a contemporary change of the journalistic culture made possible by new media technology and accelerated by the requirement to beat competitors and be first with a news item. A large majority say that they are experiencing an increased work tempo as a threat to independent journalism, which can be interpreted as a negative development of working conditions. But there are also advantages. Journalists in Russia and Poland say that they today have more time for discussion and evaluation than before, while respondents in Sweden have the opposite view. The difference here may be that Swedish journalists traditionally have had considerable space for independent work, while this space previously had been more limited in Poland and Russia.

4.4 Changes relating to ability to influence working conditions

Questions and answers about salaries and wage agreements have already been discussed in Chapter 3. Here we will only briefly highlight other general questions about the ability to influence working conditions, which was a major question addressed to the 60 journalists who were interviewed. Recent changes in the workplaces are often associated with Baumann's (2005) thesis of individualization of working life where collective solutions are replaced with individual agreements. The individual is made responsible for his or her condition in the labor market. Collective agreements are replaced by individual contracts. The change is justified by the greater flexibility but often creates an unstable ("fluid") work situation characterized by more temporary jobs than in the past, where young people in decades find themselves passed around in a carousel of temporary jobs (Deuze, 2007:9). A paradoxical phenomenon in this new work environment is the so-called out-locking in Sweden. According to the Law on Employment, jobs that last for a certain time shall automatically lead to a permanent job. But employers completely prevaricate shortly before this automatic coming into force, which means that the law protects young people from employment and not vice versa. It should be added that the uncertainty in the labor market is reinforced by the economic crisis that hit many established media companies and that often led to cuts in the number of permanent employees.

Although we made room for an elaboration of working conditions while interviewing 60 journalists, in general the interviewed did not elaborate much on the question concerning this area. Most of the Russian respondents did not discuss the question at all or limited themselves to discuss only their salaries (see Chapter 3). A summary of the interviews shows that most of the respondents answer that they have no, or limited, influence on their working conditions.

The result can be seen as strange, as most of the respondents answer "yes" to the question of their ability to influence daily work when it comes to the selection of topics and angles. But this contradiction reflects differences within the industrialized profession. Working conditions are mostly strictly regulated and difficult to influence for the individual journalist. Work content, on the other hand, is treated differently. Here the profession by tradition give space for individual initiatives (Peters and Broersma, 2012:163).

The possibilities to influence wages and daily working conditions within companies are generally more limited today, which is related to the crisis of many news companies, weaker unions and the increasing use of part-time labor. The following conversation between the interviewer and a young Polish reporter is rather typical for many of the younger reporters approached by our interviewers.

This is how a Polish journalist answers the question *Is it possible for you to have influence on your employment and your salary?*:

> "More than I have ever thought, because for quite a while I thought that I should earn more and I have been given quite a good pay rise. However, I have no influence on my employment. Because we are sort of a local branch, and the head office really decides how many people work on permanent contracts and how many on temporary ones and even my boss who would like to employ me on a permanent basis, because I am still waiting for it, I can't do that until he gets the instructions from the head office, so I have no influence." [PR2, female, 30 years old, reporter, newspaper]

A younger TV reporter gives a hint of this changing situation between generations:

> "They (the older generation, authors note) may think we're a little uncritical relating to decisions that come from above, that (…) we just bow when we get a new task. But the

organization is so slim that it is not time for us to bitch about things that one could in the good old days. Then you could have large meetings and look closely and complain about everything. Now, the union struggle is not a big deal among the young." [SR5, male, 30 years old, video reporter]

4.5 Changes in work place mobility

The concept of mobility is used here to describe the extent to which professionals may consider changing jobs or workplace.

Opinions on mobility are closely tied to what expectations you have for the future. The answers are fairly divided when it comes to viewing the issue, as evidenced by the responses to the question: *What kind of future awaits journalism as a profession in your country?*

Table 4.10: Opinions about the future of the journalistic profession (in per cent)

Country	*It will transform into another activity*	*It will vanish step by step*	*It will be the same as today*	*The profession will flourish*	*Do not know*
Poland (n 493 = 100%)	37	14	11	29	9
Russia (n 498= 100%)	33	3	34	23	7
Sweden (n 492= 100%)	28	3	31	20	16

Source: *What kind of future awaits journalism as a profession in your country due to your opinion?* The "Journalism in Change" Project Survey, 2012

Being more pessimistic concerning quality, Polish respondents also show the greatest negativity predicting the future for the profession as a whole. Over 50 per cent of the Polish respondents say that they think the profession will transform into something else or slowly vanish. But more than a third of the Polish respondents have the opposite position: it will flourish or stay the same. In Russia and Sweden the position of the respondents' answers are very alike: a majority says the profession will stay the same or even flourish.

The following conversation is maybe representative of the ambivalence that seems to characterize the answers in the survey. A young Polish TV reporter answers the question, *Are you afraid of your future?*:

"Of course I am! Each reasonable man or woman is afraid of his or her future (…). What the future will bring – this is unknown. But obviously I have a lot of confidence in that the future will be good. (…) However, how will all of this look like? God only knows." [PR 19, TV reporter]

A younger Russian TV reporter, connects quality to the social and political context, mixing despair and hope:

> "It's difficult to forecast anything for the next 5 years. Everything will depend on the socio-political situation in the country, but I don't expect major changes within the next 2–3 years. What will happen later, no one can tell. In any case, the next election cycle will be quite interesting, the old political guard are already retiring, new people are bound to come into journalism, politics, the arts. There will always be change, and it doesn't matter if it's for better or for worse, what matters is that we won't stagnate very much. In a couple of decades the country may become completely different." [RR20, female, 25 years old, reporter, television]

A middle aged Swedish magazine editor says:

> "We recently had an internal meeting on media trends. A picture of an Indiana Jones rope bridge was shown, and the editor said: "this is what we're going over, but we do not know what's on the other side." [SR 13, female, 42 years old, newspaper reporter]

The answers to the question if journalists want to stay or move within a company or move to another occupation indicate opinions about work mobility. The results from the following table are a bit surprising as it shows a strong will of continuity. The respondents seem to want to stay in a profession whose future they question.

Table 4.11: Respondents preferences on future workplace (in per cent)

Country	In the same company	In another company	With some thing else	Will be retired	Don't know	No answer
Poland (N 451 = 100%)	50	16	13	5	15	2
Russia (N 499 = 100%)	37	23	25	2	1	13
Sweden (N 467 = 100%)	47	15	10	13	13	2

Source: *Where would you like to work in five years' time?* (Only one alternative was accepted). The "Journalism in Change" Project Survey, 2012.

The most prominent desire to leave the profession is shown by Russian respondents, a position which is shared by very few respondents in Poland and Sweden. This can be seen as a reflection of the Russian phenomenon of "journalism as a social lift" (Pasti, 2014) i.e. the fact that young Russian journalists go into journalism with the desire to use it as a stepping stone to a more fashionable profession.

There are remarkably few of the interviewed who reflect on the development of a more flexible labor market. There are, though, some voices, with this Swedish example as a prominent voice. One journalist describes a labor market, which is already shifting – and of which she already is a part:

> "I see a clear trend: the editorial staff will not have large offices with 200 journalists but work with more freelancers. You have to make changes among journalists in order to match revenues. Web Magazines do not have as large a team as an editorial board which does exactly the same things every day. In one day, maybe you are working with three, in another day with 16." [RR 16, male, 38 years old, editor, newspaper]

To sum up: although there is considerable uncertainty of the future for mainstream journalism, a majority of the respondents do not state that they want to leave the business. Around half of the Polish and Swedish journalists say that they want to stay in their workplace. Russian respondents are not that stable, which probably mirrors the unstable labor market in Russia and the tendency to see journalism as a career path.

4.6 Conclusions and discussion

The first research question of this chapter was about changes in the size of the workforce and its implications for journalistic quality. A common tendency is that the size of the labor market in Poland, Russia and Sweden is declining. When asked if there has been a decrease or increase the tendency is different in the three countries, nearly 70 per cent of the respondents in Sweden say staff has been reduced in the last five years, and in Poland nearly 60 percent of journalists have the same experience. Russia is different as around a third of respondents say that they have experienced shrinking staff. But this trend has developed unevenly, where Russia as an exception has seen some increases after the recession of 2008–2009. The shrinking amount of staff in newsrooms is seen as having negative consequences for the preconditions for work in the newsroom. In the interviews we find many journalists describing increasing demands on production as a threat to quality and independence of journalism.

The second research question is about changes in skills needed in the workplace. Here journalism in most countries experienced one of the largest changes in modern times with journalists working in several different media platforms. The consequence is that one person will have to write, take pictures, use a microphone, and do camera work. The same person should be able to publish material of varying character for a web page, a printed newspaper, the radio and/or television; in other words, the journalist has to be multiskilled. Although this process has developed unevenly within the three countries, respondents perceive it as an inevitable development of the profession. The attitudes toward changes in the direction of multiskilling differ when comparing the three countries. Swedish respondents seem to be more negative toward this development, which can be seen as a consequence of the more elderly profile of the work force in Sweden.

The third research question is about the rate and volume of production, as the amount of work required from each person in a working day. The perceived tendency is the same in the three countries: the required amount of texts and programs has increased. The level of working with editing and layout has also increased in all three countries but is more prominent in Poland and Russia. Swedish journalists respond differently when asked about the space for evaluation and discussion. When working tempo increases each person not only produces more material but also does it more often and within shorter time limits. This may be considered as problematic in relation to the possibilities of verification, since there is less time for fact-checking, or at least the time perspective is shorter. About 42 per cent of Swedish respondents say that they have less space for evaluation and discussion today. On the contrary, in Poland and Russia respondents experience an increasing space for discussion and evaluation when looking back over the last 5–10 years. The difference in perception of the possibility of evaluation may be a function of different traditions within these journalistic cultures. While the Swedish journalists for a long time have worked in a liberal tradition with a more flat organizational culture, Poland and Russia have been ruled by a more authoritarian and top-down tradition.

The fourth research question is about abilities to influence the individual working conditions. The answers point in the same direction in the three states. Most of the interviewed say that they have very limited possibilities to influence these conditions in general, even if there is great freedom in choosing topics and angles in news production. This can be seen as an individualization of working life, where collective solutions are replaced with individual agreements and has to do with changes in working life in general in modern societies and especially in general and post-communist countries. The fact that outer restrictions (working contracts) are more limited, while inner restrictions are not, can be seen as a consequence of the universal artistic tradition of giving space for individual initiatives within the profession.

The final research question dealt with abilities or aspirations of moving within the profession or to another job. Even if many of the respondents are critical to the development of quality within the profession and are unsure about the future for the profession most of them want to stay within the profession, even if a small number want to change employer. This tendency is stronger in Sweden and Poland, but less visible in Russia, which can be seen as due to the younger Russian work force and the perception of the profession as a "social lift, a stage in the career to other jobs with higher status."

The survey and the interviews provide a basis for some general assumptions of a convergent development of the working conditions for journalists in Poland, Russia and Sweden, as the differences are fewer than the common traits. Firstly, the shrinking workforce in newsrooms is seen as having consequences for the quality of daily work. It results in producing working conditions considered as less favorable by most of the journalists. Secondly, the development towards multiskilling is one of the most important changes for working conditions within the profession. Today journalists to a larger extent work on many platforms as printed media, web, television and radio – at least two of these. The change is seen as inevitable for most of the respondents, even if some are critical as they perceive it will have consequences as there will be less space for specialization. Thirdly, journalists in all three states perceive that they are producing more material, more often and within shorter time-limits. Working tempo has increased, which can limit the possibilities for fact-checking and verification. Fourthly, the possibilities to influence daily working conditions within companies are more limited today, a fact which is related to the crisis of many news companies and an individualization of the bargaining process between employed and employees. Fifth, although many respondents in the survey and among the interviewed are critical of many traits of the development of working conditions, most want to stay within the profession, with the exception of younger Russian journalists, who are more open to moving to other jobs.

5. Media development and professional autonomy

Gunnar Nygren

Freedom in daily work is important for nearly all journalists in the three countries. It is the second most important factor in choosing the place of work, according to the survey. A degree of autonomy is also an important part of self-perception of journalists in their professional roles and closely related to job satisfaction (Weaver and Willnat, 2012:534).

But there are also clear differences between journalists in how big this perceived autonomy is in daily work. There are different kinds of pressure from outside and inside the media company – pressure from lack of time and formats of the media, constraints within the organization, political and commercial pressure. There is also a media development challenging the professional identity of journalists; journalistic work is undergoing fundamental changes – transitioning from a monologue to more of dialogue with the audience, with new tools and media formats introducing new kinds of expressions. Increasing commercialization and fragmentation of the media landscape are changing the foundations on which journalism rests (Deuze, 2007; Mitchelstein and Boczkowski, 2009; Singer et al., 2011).

One may assume that all these changes influence the degree of professional autonomy for journalists. But the direction is not obvious; some changes, like strong commercialization might put pressure on the degree of autonomy, but others might work in another direction; for example it is possible that technological development can both increase perceived autonomy, giving journalists more tools and a stronger position, but also limit the autonomy in an increasing demand to produce more and faster.

The purpose of this chapter is to study how perceived autonomy in daily work is related to factors on a different level, from the individual level of journalist, the position of the journalist within the media organization and to the level of media system. This analysis is also related to media development, both technological and commercial factors. From this, perceived professional autonomy is related to different kinds of media systems with various degrees of political and commercial pressure on journalism.

This allows one to raise three research questions:

RQ1: What are the differences and similarities between the three countries and media systems when it comes to professional autonomy?
RQ2: What are the most important factors explaining differences in perceived autonomy on three different levels – the individual, organizational and societal level?
RQ3: How are these factors influenced by media development? What is giving stronger or weaker perceived autonomy?

5.1 Dimensions of professional autonomy

5.1.1 Perceived autonomy in earlier research

Among U.S. journalists, perceived autonomy has been researched in surveys since 1971. Using the same questions, the results show a declining level of perceived autonomy (Weaver et al., 2007). In 1971 and 1982, 60 per cent of journalists claimed to have almost complete freedom to select stories. This figure was 34 per cent in 2013 (Weaver and Willnat 2014:11). Weaver also found clear differences between journalists: reporters in large organizations and in TV and magazines ranked lower on all indicators for perceived autonomy. A higher level of perceived autonomy was reported by beat reporters and to some extent by older reporters. There were no differences in gender, but there was a lower level among reporters from minority groups (Weaver at al., 2007:73–76). The reasons behind the decline are not clear, but Weaver can see four broad groups of limitations in the survey from 2002; pressure from source activity, professional conventions, commercial imperatives and routines/procedures in the news organization. Of these limitations, the journalists in the survey put a lot of emphasis on commercial pressure to produce more with fewer resources in combination with the influence of advertisers.

These results come from a media system with strong commercialization and a rather weak public service orientation. But what about the rest of the world? In "The Global Journalist" Weaver and Willnat (2012) take a broader perspective, comparing perceived autonomy in 22 countries. They find a strong correlation between job satisfaction and perceived autonomy. But they also find big differences between countries and no clear pattern except some Asian countries with low level of press freedom scoring low – Hong Kong, Korea and Taiwan.

In another big international survey, determinants of perceived professional autonomy are analyzed (Reich and Hanitzsch, 2013). The survey compares journalists in 18 countries, and analyze perceived autonomy in relation to other factors; power of political and economic influence, position in the newsroom, business models of the media company and level of press freedom in the country. The analysis is made on different levels – on an individual, organizational and societal level. The results show that most of the variance is explained by journalists' subjective perception of influences from political and organizational power. This is followed by objective determinants on individual and societal level. Journalists who feel limited professional autonomy, also tend to think their work is influenced by political forces (Reich and Hanitzsch, 2013:146–148). Organizational factors like commercial influence and business model show quite a low correlation to perceived autonomy. Other factors connected to media development like layoffs and multiskilling are not taken into the analysis, but Reich and Hanitzsch assume they can limit professional autonomy.

5.1.2 Perceived autonomy in three questions

The analysis in this chapter is based both on the survey and on interviews with 60 journalists in Poland, Russia and Sweden. In the survey, the starting point is three questions concerning perceived autonomy in the daily work. All three questions have also been used by Weaver in the survey of the U.S. journalists (2007) and in some of the surveys in "The Global Journalist" (Weaver and Willnat, 2012). The results show a strong correlation between the three questions in the patterns of the answers (Pearson r=0.44–0.65). Together they give a general picture of the perceived autonomy in daily decisions in the three countries:

- Nearly half of the journalists say they almost always are able to get a subject covered. Only a few answer only occasionally.
- 28–29 per cent of journalists say they have almost complete freedom to select stories and which aspects of the story that should be emphasized. Half say "a great deal", and only 17–18 per cent say they have only some freedom in these aspects.

The level of perceived freedom is almost the same as in the recent U.S. survey, where 52 per cent say they "almost always" can get a subject covered. But the share saying "almost complete freedom" to select stories and to decide the emphasis is clearly lower among Swedish, Polish and Russian journalists, than among their U.S. colleagues; 28–29 per cent compared with 40 and 42 per cent in the U.S. survey. The European journalists in this survey seems to have an equal feeling of having an influence on news evaluation as a group, but not on an individual level (see Table 5.1).

Table 5.1: Questions on perceived autonomy in daily work (in per cent)

If you have a good idea for a subject which you think is important and should be followed up, how often are you able to get the subject covered?	I don't make such proposals	Only occasionally	More often than not	Almost always	Do not know	N
	3	13	34	47	3	1487
How much freedom do you have….	No freedom at all	Only some freedom	A great deal of freedom	Almost complete freedom	Do not know	N
…in selecting the stories you work on?	4	18	47	28	3	1483
…in deciding which aspects of a story should be emphasized?	2	17	49	29	3	1478

Source: The "Journalism in Change" Project Survey, 2012.

These questions have a clear correlation with the evaluation of freedom in work in general in the workplace. Those journalists who think that freedom in work is important, also report a higher degree of individual freedom to select stories and emphasis in their stories (r=0.24 and 0.19). If you think that freedom in work is important, you also look for workplaces able to offer you a large degree of autonomy in daily work.

This is the average result of the survey. A closer analysis tells us more about differences between groups of journalists, about what kind of indicators are correlated to higher or lower levels of perceived autonomy. In so far as we can identify these indicators, we can also discuss the consequences of autonomy in daily work in media development.

5.2 The individual level

There are clear correlations between professional experience, age and gender in the perceived autonomy. The longer experience you have as a journalist, the more you feel autonomous in the daily decisions (r=0.130–0.146). Some examples: Among journalists with 26–30 years of experience, 41 per cent say they have almost complete freedom to select stories and to decide emphasis. Among young journalists with less than five years of experience the same figures are 20 per cent and 23 per cent.

This is also visible in age groups; among young journalists under 35 years old the level of perceived autonomy is clearly lower than among older and more experienced colleagues. This difference increases as we consider also the factor of gender – among female journalists under 35 years, 23 per cent claim they have almost complete freedom to select subjects compared with 37 per cent among male journalists over 50 years old (see Table 5.2).

The differences between male and female journalists are generally equal or larger than between age groups. This can be a result of both differences in factual autonomy and also differences in how the individual male or female journalists perceive their influence in daily work; it can be a result of men usually estimating their influence higher than women. There are no patterns in the differences between the countries in the gap in perceived influence between men and women; in all three countries men report a higher degree of perceived autonomy than women with differences up to 11–12 per cent units. A comparison: among U.S. journalists the differences were very small and even in favor of women in the survey 2002 (Weaver et al., 2007:76).

Table 5.2: Perceived autonomy in age and gender (in per cent)

	Under 35 years old		36–50 years old		Over 51years old		All ages	
	Male	Female	Male	Female	Male	Female	Male	Female
Almost always can get a subject covered	50	42	52	51	58	46	53	46
Almost complete freedom in selecting stories	31	23	36	26	37	27	34	25
Almost complete freedom in deciding emphasis	27	26	35	25	37	33	33	27
N=	250	399	243	212	166	118	659	729

Source: The "Journalism in Change" Project Survey, 2012.

The position in the newsroom is important for perceived autonomy. The editors, the editorial writers and columnists have a larger perceived autonomy than average, they have a position to choose subjects and emphasis. On the lower part of the list we find subeditors/layout journalists, reporters and online editors with a clearly lower perceived influence in daily work (Table 5.3).

Table 5.3: Perceived freedom for different groups in the newsroom (percentage of each group)

	Almost complete freedom in selecting stories	*Almost complete freedom in deciding emphasis*	N
Reporters	29	27	777
Editors	36	40	405
Subeditors/layout	23	24	195
Photographers	35	34	116
Commentators	37	41	178
Editorial writers/ columnists	35	33	101
Producers in radio/TV	31	44	75
Online editors	26	26	182

Source: *Some questions about your daily work: How much freedom do you have in selecting the stories you work on? How much freedom do you usually have in deciding which aspects of a story should be emphasized?* The "Journalism in Change" Project Survey, 2012.

In the U.S. survey, beat reporters had a larger perceived autonomy. This is not confirmed in the present survey, the difference is only one or two per cent, and not significant. However there are some differences between beats – reporters covering crime, business and society have in general a higher influence to get a subject covered. Reporters covering foreign affairs, human interest stories, and lifestyle have clearly a lower influence.

Different kinds of temporary employment are becoming more usual in media business (Deuze, 2007:147). This is also the fact in journalism, but still the majority have a steady job. About 71 per cent of all journalists in the survey have a steady job, 14 per cent are freelancers and 15 per cent have different kinds of temporary employments. There are clear differences between the groups when it comes to perceived autonomy. Journalists with temporary jobs have clearly lower levels of perceived autonomy than those with steady jobs. Freelancers are quite free to choose the subject, but have fewer possibilities to get a subject covered (See Table 5.4).

Table 5.4: Employment and perceived autonomy (percentage in each group)

	Almost always can get a subject covered	*Almost complete freedom in selecting stories*	*Almost complete freedom in deciding emphasis*	N
Steady job	52	28	30	1022
Temporary job	32	16	20	162
Freelancer	43	41	35	208
Other	57	41	24	51

Source: *If you have a good idea for a subject which you think is important and should be followed up, how often are you able to get the subject covered? How much freedom do you have in selecting the stories you work on? How much freedom do you usually have in deciding which aspects of a story should be emphasized?* The "Journalism in Change" Project Survey, 2012.

Also incomes can influence perceived autonomy. In Sweden over 90 per cent of journalists say they have enough income from journalism to provide for all necessary expenses, and most of them are also able to afford some extra things. In Poland and Russia the same figure is much lower: only between 51–57 per cent of journalists have a good income from journalism (see also Chapter 3.4). In Poland, there is a clear correlation between income and perceived autonomy – those journalists having a good income?? also feel a larger degree of freedom in daily work. The correlation is $r=0.124$ in selecting stories and $r=0.202$ in deciding emphasis. This shows a stronger economic pressure on journalists in Poland than in Sweden and Russia.

5.2.1 Freedom within given frames

In almost all interviews, journalists describe a large space for their own initiatives. But at the same time they describe clear limits for what are possible, frames to be aware of in daily creative work. These frames can be limited resources and limited time, formats and editorial policies and power structures within the media company controlling the work of journalists.

Experienced journalists, often with a special area to cover, give a picture of great independence in daily work. A Swedish reporter on a special magazine for women tells about her area of testing new products: "I have my niche, I am employed to handle this area and I have my expertise" [SR10]. Also experienced reporters in Poland and Russia emphasize that their position in the newsroom gives them a large amount of autonomy:

> "I have worked there long enough and editors know me well enough so I do not have to consult each step with them while preparing my reports." [PR8, female, 35 – 55 years old TV reporter]

> "Now I am a quite well known person at the agency, so I have huge space for initiatives and I offer them quite often. In general we have a creative atmosphere, any ideas and offers of my colleagues are mostly accepted as well." [RR6, male, 44 years old, editor, news agency]

Given formats and the editorial policies of media companies are clear limits for the initiatives of individual journalists. Local journalists have to cover the local area; niche magazines have to focus on subjects relevant for a niche audience. A Swedish journalist working for a magazine for members of a trade union emphasizes this niche:

> "(…) but within these frames we have a quite large space… When I was a freelance, I did not have these kind of frames… so it is more limited now." [SR14, female, 37 years old, reporter, magazine]

Also Polish and Russian journalists talk about these frames given by media format. A Polish web editor describes his daily work as giving a reflection of what is produced for the radio station connected to the news site. But he can try to convince his colleagues on radio to cover some subjects or sometimes do it himself [PR16]. A Russian TV-journalist is very clear in describing this relationship between the individual journalist and the media format:

> "A journalist should always have the initiative, but there are different formats. For example, now we are preparing materials exclusively for Moscow and within that framework I suggest my own ideas. The same goes for political news…. Initiative from journalists always comes within the established boundaries." [RR14, male, 37 years old, TV journalist]

Swedish journalists often talk about lack of time and resources as a limit for their own initiatives:

> "There is a large space for our own ideas. The greatest obstacle is the lack of time." [SR9, female, 40 years old, TV reporter]

Many reporters tell about how the work with their own projects beside their daily work, when they have some time left in the daily production. One freelance reporter says that even if she is hired to work for a newspaper like Metro, she has to work with longer projects besides the short stories for the daily production [SR18]. Another reporter gives a similar picture:

> "The problem is to have enough time. With high pressure during working time there is no space in the organization for your own initiatives. I have to do this in extra time or in my free time, just because you enjoy it. Take one evening without pay. The newspapers receive the ideas, but have no working time to give… " [SR1, male, reporter, 56 years old, a daily metropolitan newspaper]

The same reporter also describes the news organization as more of a top-down control. The individual journalist does not have the free will to say what he wants today – "You are more careful today, more concerned about your position."

Polish journalists describe a large space for their own initiatives and ideas in the interviews. The most common limit is if the ideas are accepted by colleagues and in meetings in the newsroom. One male radio editor on a regional radio station says there is "full autonomy" – but what they are finally producing is what is discussed by the daily meeting. Ideas are discussed and changed during the process – "so it is more about shaping the information, not just an automatic choice" [PR12].

Also Swedish and Russian journalists describe how ideas and initiatives are developed during meetings and discussions in the newsrooms and with editors. But in the Russian interviews, the journalists more often tell about limits and unwritten rules in the newsroom. A One of them gives this answer on the question about personal ideas:

> "It depends on what is proposed. Our suggestions are rarely heard, we normally follow whatever we are told by our superiors. On the whole we just make news, we have a list of forbidden and recommended aspects that we try to bear in mind when we prepare material. [RR4, male, 25 years old, radio, sports reporter]

Another journalist talks about freedom for initiates, as long as they are within the profile of "military patriotic subjects" of the channel:

> "Light-hearted topics do not always make the cut, but initiatives usually come from us. We don't have censorship in this respect, there is quite a lot of freedom, provided the journalist keeps in line with the specifics of the channel." [RR11, male, 24 years old national TV channel, producer]

A journalist worked in a small metropolitan TV channel. Dozhd gives a similar description of quite narrow frames for daily work, saying initiatives are "relative":

> "There are let's say, 10 topics and we work on them, that is our agenda. On the other hand, if I learn something interesting I can suggest it and we go out and check it. …On the whole, news journalism gives little room for creativity. It is like pancakes, you do it well but it is nothing extraordinary, everything falls within the present limits…" [RR20, female, 25 years old, small metropolitan TV channel]

Some old Russian journalists talk about very narrow limits for what is possible to influence from the individual journalist. One of them, who works far away from Moscow, describes it like this:

> "Now looking with the wisdom of years passed in the profession, I understand myself as young; because the inner censor sitting in me is so strong that any ObLit[1] is nothing in comparison. The internal censor is the 8th or 9th sense of self-preservation acquired over the years in the editor's office. It is something all journalists have and it only grows stronger over the years." [RR3, female, 62 years old, newspaper, editor, chair of local department of Union of journalists]

Common for all journalists in the interviews is a feeling of having a clear space for their own decisions. But it is autonomy "within frames" of what is decided in the media company – format, time, editorial policy and what the audience expects. To produce content every day, it is necessary for the media company to give journalists autonomy within these frames. But most journalists are also very much aware of the limits of their professional autonomy at the individual level.

5.2.2 Professional identity and autonomy

The factors connected to professional identity and professional commitment can also be important for perceived autonomy. But the survey shows diverging results in these areas. Having a professional education or not makes no difference for perceived autonomy. The levels are equal in all groups regarding education. If you belong to a professional union or not, it makes no difference either – the perceived autonomy in daily work is the same whether you are a member of the union or not.

There is a clear correlation between putting pride in being a journalist, and the feeling of autonomy in daily decisions. The journalists in the survey evaluated the statement "I am proud to call myself a journalist", and among those who fully agree, 36 per cent also say they have almost complete freedom to choose stories to work on. Among those giving a neutral answer to the statement on pride, only 19 per cent claimed to have almost complete freedom to choose stories. The correlation between pride and choosing stories is r=0.159 and between pride and deciding emphasis r=0.176.

There is also a clear correlation between future plans and the perceived autonomy in daily work. Those who think they will leave journalism within five years, report lower levels of perceived autonomy. Among those planning to leave journalism, 40 per cent say they can get a subject covered. The same figure among those who would like to stay in the same media company is 54 per cent (see Table 5.6). The questions on pride and on future plans show a clear connection between professional commitment and perceived autonomy, and this perceived influence in daily work seems to be an important part in professional identity.

---

1 The state authority carrying out censorship in Soviet times.

Table 5. 5: Future plans and perceived autonomy (percentage in each group)

Where would you like to work in five years from now?	*Almost always can get a subject covered*	*Almost complete freedom in selecting stories*	*Almost complete freedom in deciding emphasis*	N
In the same media company	54	32	33	653
In another media company	46	22	23	258
In something else than journalism	40	28	28	232
I will be retired	37	30	28	93
Don't know	48	28	25	133

Source: *Where would you like to work in five years from now? If you have a good idea for a subject which you think is important and should be followed up, how often are you able to get the subject covered? How much freedom do you have in selecting the stories you work on? How much freedom do you usually have in deciding which aspects of a story should be emphasized?* The "Journalism in Change" Project Survey, 2012.

Relation to the political system is a key dimension in studies of autonomy for journalism, especially in post-communist countries (Roudakova, 2012:253; Vartanova, 2012:141). But on the individual level, the study shows very few correlations between political attitudes and the level of perceived autonomy in daily work. There is no correlation at all among journalists in any of the countries between political values (if you consider yourself left or right) and autonomy in daily work. There is no correlation either between an alternative for you to be active in politics and perceived autonomy. Only on the question about the level of political interest, are there clear correlations with perceived autonomy mostly among Russian journalists (see Table 5.6). Those Russian journalists who are interested in politics also report a higher degree of autonomy in daily work. This says nothing about whether these journalists agree or not with the politics of the current leaders, it is just that those who are interested in politics also feel they can create more space for their own decisions in daily work. Or perhaps they just think this space is important for them.

Table 5.6: Correlation between interest in politics and political questions in society and... (Pearson correlation, only significant correlations are given)

	How often are you able to get a subject covered?	How much freedom do you have in selecting stories?	How much freedom do you have in deciding emphasis in the story?
Poland	–	–	0.138** N=475
Russia	0.216** N=475	0.238** N=482	0.188** N=477
Sweden	–	0.090* N=477	–

*** Correlation is significant at the 0.01 level (2-tailed)*
** Correlation is significant at the 0.05 level (2-tailed)*
Source: The "Journalism in Change" Project Survey, 2012.

5.2.3 Individual or collective autonomy?

A certain degree of autonomy in work is an important part of the professional culture for journalists, a part of their professional pride. When the journalists in interviews talk about their daily work, most of them talk about their independence even if they also are aware of the limits. This autonomous space within the limits and formats of the media is a key component of the professional culture.

During a process of professionalization, an important brick is the building of professional institutions. One of these institutions is professional trade unions, giving journalists a collective strength to defend professional standards and autonomy (Engblom 2001:260). Among Polish and Russian journalists, only a minority of journalists are members in any union, 23 and 29 per cent. Most of the members are old, and the share among young journalists is even smaller. Unions are regarded to be part of the old system, and one Polish journalist explains his opinion:

> "For me, journalism means independence and if I ever felt in danger I know that no trade union, absolutely no trade union would help me and somewhere someone can reproach me for belonging to some unions." [PR1, male, under 35 years old, radio, reporter]

In Sweden, over 80 per cent of journalists are members of the Union of journalists. But the same trend is visible in Sweden – two of the young journalists in the interviews have left the union because they are disappointed by it. Another young journalist active in the local board of the union says:

> "(...) the trade union struggle is not a big thing among young journalists." [SR5, male, 30 years old, TV, video journalist]

In his company it is difficult to find members for the local union and to find time for meetings.

Autonomy for journalism as a social field is important for journalists in all three countries. But the emphasis and kind of obstacles differs clearly. In Sweden, journalists in the interviews talk much about their role as watchdogs in society, the journalistic profession is needed with their knowledge to scrutinize power in society. This is a role that cannot be taken by bloggers or social media.

A reporter in regional TV news also emphasizes the special knowledge of journalists:

> "In blogs you can get opinions. But our task is to inform, explain and to question things. This is something our occupation needs to talk more about in the future, which separates us from other groups... Verification will be very important for both journalists and media consumers in the future." [SR9, female, 40 years old, TV, reporter]

In Poland and Russia, journalists in the interviews also talk about the borders of journalism, but not in relation to bloggers. Instead the emphasis is PR and others using journalism for their own purposes. A Polish reporter on a weekly political magazine emphasizes he wants to protect the borders of the profession:

> "(...) Some important elements of journalistic ethics so this profession could protect its margins of freedom...so that the role of a journalist won't be confused with the role of a PR worker, the role of an advertising agent, so that people have a clear view of the border between the groups of people who write in good faith, sometimes with subjective views, but not for money." [PR4, male, 35–55 years old, magazine, reporter]

Other journalists talk about a crisis of values among Polish journalists and the need to get back to a "social responsibility" [PR7]. A veteran Polish TV reporter, 76 years old with experience from working in the U.S., describes Polish journalism as a "simulation of journalism" – there is a lack of both ethics and a lack of verification in Polish journalism, he says:

> "(...) it seems to me that there is very little awareness of the critical situation for the profession and the media are in and that people are mainly concentrated on distinguishing a right-wing journalist from a left-wing one and too little attention is paid to fundamental issues of the profession like whether the journalist is provided with the right conditions to inform the public or not." [PR6, male, over 70 years old, TV, commentator]

In the interviews, Russian journalists seem to be very polarized in their evaluation of development. One of them is very optimistic. He has found a kind of autonomy in his project publishing journalism in a blog, and being supported by the readers with donations:

> "I have created my own information channel within the blog-project, money paid by readers can pay for a living if you work and write well. My constant work shows that a journalist can work in a way that makes an editor's office unnecessary ..." [RR1, male, 35 years old, freelancer]

Other Russian journalists within traditional media companies also report a change, but not in the same positive manner. Some of them see paradoxes:

> "Today everything is fine, no one throws journalists in jail, and no one shuts anyone's mouth although many believe there are many journalists for sale in the industry. That is wrong. There are journalists who make money from trifles. There is no global corruption here. People on national channels have highly developed self-censorship, you have to think twice before you say something." [RR14, male, 37 years old, TV, correspondent]

Another print journalist in Moscow can see a development from journalism in the 1990's when everything was permitted and journalists took the risk of being beaten up:

> "Now we see mainly sponsored, purchased views... Now many mass media are restrained perhaps not by censorship but self-censorship. Investigative journalism is not taken seriously at all; it is all thought to be a purchased opinion, a sponsored journalism." [RR16, 40–45 years old, newspaper, journalist]

Other Russian journalists talk about the profession as not taking responsibility for their work [RR8]. They describe freedom as opposite in relation to responsibility:

> "We wanted freedom, but I see it, we bit off so much of it that we could not chew and now we are suffering.... We have to retreat, move back to moral values away from the cult of money, cult of latest intelligence, cult of journalistic recklessness." [RR12, female, 28 years old, weekly magazine, editor]

There is a kind of double vision on professional autonomy among journalists in all three countries. In all countries, a degree of autonomy is an ideal. But at the same time there is a gap between this ideal and reality, giving constraints. In Sweden, journalists are mostly talking about conditions in daily work as an obstacle to this autonomy. In Poland and Russia, the gap seems more complicated in terms of both public relations and the responsibility to society.

At the same time, journalists in all three countries can imagine themselves changing jobs, to start working in public relations and communication. In that case they would not need to struggle with this gap between ideals of autonomy and reality anymore. A young Swedish reporter on a daily newspaper says she sees how journalism is being reduced, colleagues have to leave. If there is not time to do a good job, "I will take my education in economics and do something else," she says [SR4]. Also young journalists in Poland talk about going to another kind of job – as a spokesperson or in PR.

To sum up, the indicators on a personal level show a clear connection between perceived autonomy in daily work and professional identity. Long experience, a steady job, pride over being a journalist and the plans to stay in the media company indicate also a high level of autonomy. The negative indicators are connected to the opposite – less experience, temporary employment, low incomes and a subordinate position in the newsroom. Political factors seem generally to have a small role at the individual level. But another factor gives a clear difference in the figures from the survey; gender. Female journalists in all ages in the three countries report a lower level of perceived autonomy in all three questions.

In the interviews journalists talk about a large space for their own initiatives, within the frames given by the media organization. To handle this space and keeping some kind of professional pride is a daily task for the individual journalist.

5.3 Organizational level

At the organizational level, factors like business models and work practices might have an influence on perceived autonomy in the daily work of journalists. Also changes in daily work like new technology and downsizing can have an impact.

The journalists in the survey have been placed in five groups according to the business model of the media company where they work: Printed paid newspapers, magazines and weeklies, state and license funded radio and TV, commercial ad-paid media (TV, radio, free sheets and online) and journalists working as content producers outside the newsrooms (news agencies, freelancers and production companies).

There are clear differences in the perceived autonomy between these groups (see Table 5.7). The general level differs between the countries – the figures in Poland are high, in Sweden clearly lower and in Russia even lower. But the patterns are the same comparing the groups of journalists: In printed paid newspapers, perceived autonomy is high, they have many experienced journalists and established practices giving journalists much space for their own decisions. In magazines and ad-paid commercial media the level is a bit lower, here the commercial pressure can be stronger. In public or state funded radio and TV the level of autonomy is the lowest, especially in Russia, where journalists in state radio/TV report the lowest figure of all, on their ability to get a subject covered. Also among subcontractors the level is low, they have to listen to the customers of their products giving the journalists less influence.

Table 5.7: I can almost always get the subject covered *(percentage in each group)*

	Poland	Russia	Sweden	All countries	N
Printed paid newspapers	73	47	51	55	496
Magazines	54	41	44	47	329
Public service/state radio and TV	48	26	44	42	227
Commercial ad-paid media*	60	44	42	50	277
Subcontractors**	69	32	34	37	118
All groups	59	42	45	49	1447

* commercial TV and radio, free newspapers and online only)

** news agencies, production companies and freelancers

Source: *If you have a good idea for a subject which you think is important and should be followed up, how often are you able to get the subject covered?* The "Journalism in Change" Project Survey, 2012.

There can be different reasons behind the differences:

- Less freedom for journalists in public service/state TV and radio in general can be a consequence of TV as a technical platform giving the individual journalist less freedom. TV is more teamwork, giving less autonomy to the individual journalist, and in the U.S. survey Weaver also reports a lower degree of perceived autonomy in TV (Weaver, 2007:74).
- There are also clear differences between countries. In Russia and Poland, journalists in commercial media report a higher degree of autonomy than in state controlled TV and radio, while the level is nearly the same in Sweden. This indicates there is an element of political control in the difference, more tight limits for journalists in state/public service media than in the commercial media sector.

In the U.S. survey from 2002, the size of the newsroom was an important indicator of perceived autonomy in daily work. In small newsrooms journalists had clearly a greater degree of autonomy than in big newsrooms (Weaver, 2007:74). This is not the case in the present survey, for the differences are in general very small between journalists in small and big newsrooms. But another factor has a clear connection – if there has been downsizing in the newsroom in the last five years. In Sweden and Poland, most journalists have experienced the staffing in the newsroom to be shrinking some or much in the last five years – 69 per cent in Sweden and 58 per cent in Poland. In Russia the development is more stable, 32 per cent report a shrinking newsroom, and 25 per cent report increased staff.

This development shows a clear correlation to the perceived autonomy (see Table 5.8). Journalists working in shrinking newsrooms report lower levels when it comes to the freedom to get a subject covered, than colleagues working in an environment more stable or expanding (r=0.14). Journalists working in shrinking newsrooms have less freedom to choose subjects and angles, than journalists in stable or expanding newsrooms. The difference is 10–17 per cent units, and shows a harder pressure on journalists still keeping their work in downsized newsrooms.

Table 5.8: Perceived autonomy in shrinking newsrooms (percentage in each group)

Has the staffing in the newsroom increased or decreased the last five years?	*Almost always can get a subject covered*	*Almost complete freedom in selecting stories*	*Almost complete freedom in deciding emphasis in the story*	N
Decreased much	44	25	25	408
Decreased some	57	33	30	285
No change	62	28	29	220
Increased some	49	29	36	170
Increased much	58	35	42	62

Source: *Over the last five years, has the size of your editorial staff grown or shrunk? If you have a good idea for a subject which you think is important and should be followed up, how often are you able to get the subject covered? Some questions about your daily work: How much freedom do you have in selecting the stories you work on? How much freedom do you usually have in deciding which aspects of a story should be emphasized?* The "Journalism in Change" Project Survey, 2012.

Daily work in newsrooms has changed over the last 5–10 years, and journalists report direct production to take much more time (writing, editing, producing). The higher level of production shows no correlation to autonomy in daily work. The only part of the work process having a clear correlation to perceived autonomy is time for research and verification. Those having more time for research, also report a higher degree of perceived autonomy: among those having more time for research 43 per cent report an almost complete freedom to decide the emphasis in their story. The figure among those journalists spending much less time for research is 27 per cent having almost complete freedom (see Table 5.9).

Table 5.9: Perceived autonomy in daily work in relation to time spent on research and verification (among those with 5–10 years of experience) (percentage in each group)

	Almost always can get a subject covered	*Almost complete freedom in selecting stories*	*Almost complete freedom in deciding emphasis*	N
Much less time	43	31	27	84
Less time	44	32	30	175
The same time	50	27	29	360
More time	47	24	26	297
Much more time	60	39	43	275

Source: *If you have a good idea for a subject which you think is important and should be followed up, how often are you able to get the subject covered? Some questions about your daily work: How much freedom do you have in selecting the stories you work on? How much freedom do you usually have in deciding which aspects of a story should be emphasized?* The "Journalism in Change" Project Survey, 2012.

There is a very clear correlation between freedom to manage your own time and if you feel you can choose subjects and decide on the emphasis (r=0.495 and r=0.330). For example:

- Among those having complete freedom to manage their own time, 56 per cent also say they have complete freedom to choose the subject.
- Among those having only some freedom to manage their own time, only 11 per cent say they have complete freedom to choose the subject.

This shows a strong connection between more general freedom in daily work and perceived autonomy in content production.

5.3.1 Technical development and perceived autonomy

Other changes in daily work are related to technological development, both in the tools of daily production and in an increasing interactivity in relation to the audience. Analysis shows some of these changes to be positively related to perceived autonomy in daily work. One is the demand on journalists to be multiskilled – to be able to work with all parts of the production process and on different platforms.

Journalists saying they feel like a multi-reporter, they also say they have a greater degree of freedom at work (see Table 5.10); they can get subjects covered, they can choose stories and decide on the angle (r=0.112–0.176). Journalists who agree that multiskilling makes journalism more creative and puts more power in the hands of the individual journalist, also report a higher level of perceived autonomy. The pattern is clear in all countries, but it is stronger in Russia where the correlation is r=0.18–0.21. Some examples:

- Among those feeling as a multiskilled reporter, 57 per cent say they almost always can get a subject covered. The figure among those not describing themselves as multiskilled is 38 per cent.
- Among those who agree that multiskilling gives more space for creativity, 33 per cent have almost complete freedom to choose the subject. The same figure among those who do not agree is 24 per cent.

Table 5.10: Correlations between attitudes towards multiskilling and perceived autonomy (Pearson correlation, only significant correlations are given)

	How often are you able to get a subject covered?	*How much freedom do you have in selecting stories?*	*How much freedom do you have to decide emphasis in stories?*
I feel like a multiskilled reporter	0.176** n=1373	O.168** n= 1385	0.112** n=1381
Journalists in my media organization are expected to be multiskilled	0.212** n=1401	0.122** n=1410	0.097** n=1407
Multiskilling means more space for creativity	0.157** n=1393	0.119** n=1403	0.112** n=1399
Multiskilling will decrease the quality of journalism	–	–	–
Multiskilling gives more power to the individual journalist	0.108** n=1371	O.101** n=1379	0.063* n=1374

*** Correlation is significant at the 0.01 level (2-tailed)*
** Correlation is significant at the 0.05 level (2-tailed)*
Source: The "Journalism in Change" Project Survey, 2012.

Another big change is the increasing interaction with the audience through different channels, from traditional channels like the telephone, to comments on web and social media like Facebook and Twitter. In the survey journalists have answered questions on how they interact with the audience, and there are no correlations at all between what channels you choose for interaction and perceived autonomy. If you use telephone or Twitter, it does not matter in relation to your perceived freedom to choose subjects and angles.

There are some clear correlations when it comes to attitudes towards interactivity. Those journalists with a positive attitude towards listening to the audience report a slightly higher degree of autonomy in their daily work. Also those who believe increased interactivity gives higher quality in journalism, feel themselves to have a higher degree of autonomy in daily work. Among those who believe in increased quality with interactivity, 33 per cent say they can choose their own stories. The same figure among those negative to the impact on quality is only 23 per cent.

This means that a journalist who wants to listen to the audience, also reports a slightly higher degree of autonomy in decisions in daily work. This is also confirmed in other parts of the survey where the influence of journalists and of the audience on content goes hand in hand (see Table 5.13), at least in the eyes of journalists, who see the audience as an ally in the struggle for their own influence. To some extent, this can be a result of the age factor – young journalists are in general more positive towards multiskilling and towards interactivity. This gives them also a higher self-confidence in the perceived autonomy in daily work.

To sum up: On the level of organization and work practices, there are clear differences in the perceived autonomy in daily work. Not in relation to the size of the workplace, but if the staffing is shrinking or not; shrinking staff in the newsroom gives less freedom to decide subjects and angles. Freedom to manage your own time and to have more time for research and verification is positively related to perceived autonomy in decisions. New practices with new technology are positively related to perceived autonomy, according to the survey. Journalists with experience and positive attitudes towards multiskilling and increasing interactivity, also report higher levels of autonomy in daily work.

5.4 Societal level

The division between factors on the organizational and societal level is difficult to make. Media organizations and different kinds of actors outside media are intertwined to various degrees in the three countries. One part of this mixture is what Hallin and Mancini (2004), label as "political parallelism" when media and politics are part of the same system through ownership, the use of media for political purposes, and media people go in and out in politics. The degree of political parallelism is one of the dimensions used by Hallin and Mancini to define different media systems. In their analysis Sweden is regarded to have a low degree of this, and belong to what they call a "democratic corporative system". Poland and Russia are not included in their first analysis, but in a follow up in 2012, researchers find a high degree of political parallelism in these countries (Dobek-Ostrowska 2012:37–38; Vartanova 2012:128–131).

It is not only the borders between media and politics that are unclear. In an increased commercial media sector, the borders in relation to business interests and advertisers are both outside and within media companies. One important part of professionalization has been to have a "wall" between newsrooms and commercial interests of the media company (Waisbord, 2013: 58). This division between journalism and commercial interests is questioned when media companies are under severe economic pressure, and the position of advertisers and ad-departments within media companies grows stronger. It also gives the owners strong motives to interfere more actively in the running of the media company.

So the analysis of perceived autonomy in relation to factors on a societal level becomes a mix of variables inside and outside media companies. Political connections inside the company and political pressure from outside go hand in hand. Commercial pressure comes from both other parts of the media company (market- and ad-departments), from owners and from advertisers/business outside the company.

On the level of countries, differences in perceived autonomy among journalists are relatively small (see Table 5.11). Journalists in Poland report a higher level of being able to get a subject covered, and Russian journalists a lower level on choosing stories to cover. But these differences are not big, and they are also possible to explain with other factors like the young work force in Russia. It seems like a certain degree of autonomy in the daily decisions in work is a common trait of the journalistic profession, independent of the definition of media system in the country.

Table 5.11: Perceived autonomy in daily work in Poland, Russia and Sweden (percentage in each group)

	Almost always can get a subject covered	*Almost complete freedom in selecting stories*	*Almost complete freedom in deciding emphasis*	N
Poland	59	31	30	490
Russia	42	25	30	487
Sweden	45	31	30	473

Source: The "Journalism in Change" Project Survey, 2012.

5.4.1 Influence of other groups

However, there are clear differences looking deeper into the question of the influence on content. When journalists in the survey evaluate the influence on the choice of subjects of different groups there are both similarities and differences (see Table 5.12) The most important factors are the same in the three countries:

audience interest, professional interest of the journalist and the editorial policy of the media company. There are also important differences:

- The political affiliation of the media company and strong political actors outside the media are regarded as being much more important by journalists in Russia, and to some extent in Poland, than in Sweden.
- The influence of advertisers is regarded as being much more important in Poland and Russia than in Sweden.
- In Sweden, influence from political actors inside and outside the media company is almost non-existent. Advertisers and businesses outside the media company are regarded as having some influence, but not as much as in Russia and Poland.

Table 5.12: Different factors influencing the selection of subjects in daily work (Mean on a scale 1–5 where 5 is very significant)

	Poland			Russia			Sweden		
	M	SD	N	M	SD	N	M	SD	N
Professional interest of journalist	4.03	1.04	484	3.89	0.98	492	3.96	0.89	467
The editorial policy of the media company	3.89	0.99	474	4.21	1.87	493	3.56	1.01	455
The political connection (affiliation) of the media company	2.78	1.31	415	3.26	1.25	467	1.58	0.95	455
Strong political actors outside the media company	2.44	1.27	382	2.79	1.23	447	1.82	1.00	445
Audience interest	4.29	0.88	481	4.02	0.96	494	4.05	0.80	477
Needs and interests of society	4.12	1.01	470	3.68	1.08	479	3.48	0.98	468
Those who advertise in the media	3.30	1.35	380	2.95	1.31	439	1.87	1.20	451
Business interests outside the media company	2.80	1.30	382	2.52	1.27	425	2.00	1.19	441

Source: *Different factors influence the selection of subjects in the daily work of your media organization. How would you evaluate the significance of the following factors on your work in your news organization?(M=mean, SD=standard deviation, N=number of answers)*The "Journalism in Change" Project Survey, 2012.

This general evaluation of influence in the newsroom correlates clearly with the perceived influence of the individual journalist in daily work (see Table 5.13).

Not surprisingly, there is a strong positive correlation between professional interest of the journalist in selecting subjects and the autonomy in daily work, especially in the choice of stories to work on. There is also a clear positive correlation to influence of the audience. This audience influence can be seen as a market influence, a pressure to adapt content to the market. But the positive correlation with the influence of journalists indicates that the audience can be used by journalists to increase their own influence.

There are also clear negative correlations between other kinds of influence and autonomy of journalists in daily work. The strongest negative correlation is in relation to political influence, both when it comes to political connection of the media company and political actors outside the media company trying to influence selection of subjects. This negative correlation is especially strong among Russian journalists. Those journalists who feel a strong political influence in the selection of subjects, also report lower levels of their own autonomy in daily work. This negative impact of political influence is clearly stronger than the negative impact of commercial actors outside the media company, from advertisers and business outside the media company. This is more the case in Poland and Sweden, than in Russia.

Table 5.13: Correlations between the influence on selection of subjects and perceived autonomy in the daily work (Pearson correlation, only significant correlations are given)

Factors influencing selection of subjects in the newsroom:	*How often are you able to get a subject covered?*	*How much freedom do you have in selecting stories?*	*How much freedom do you have to decide emphasis in stories?*
Professional interest of journalist	0.224** N=1404	0.289** N=1414	0.260** N=1409
The editorial policy of the media company	–	–	–
The political connection (affiliation) of the media company	-0.124** N=1300	-0.151** N=1310	-0.167** N=1307
Strong political actors outside the media company	-0.131** N=1327	-0.185** N=1248	-0.177** N=1245
Audience interest	0.124** N=1408	0.158** N=1420	0.120** N=1418
Those who advertise in the media	-0.104** N=1235	-0.092** N=1246	-0.118** N=1241
Economic power outside the media company (big companies, etc.)	-0.086** N=1211	-0.098** N=1224	-0.144** N=1217

** *Correlation is significant at the 0.01 level (2-tailed)*
* *Correlation is significant at the 0.05 level (2-tailed)*
Source: The "Journalism in Change" Project Survey, 2012.

When journalists in the survey evaluate how the influence on media content has changed over the last 5–10 years, the influence of audiences is becoming much stronger in all three countries (see Table 5.14). This can be a sign of the effects of increasing interactivity in a network society, as well as a sign of increasing market competition and "market-driven journalism". Other groups in the media company (departments for market and ads) have also clearly increased their influence in all three countries.

However, there are also important differences in the changing influence, as perceived by the journalists. Differences are sometimes more difficult to explain:

- In Poland, journalists are experiencing strong commercialization. The influence of advertisers and owners is increasing sharply. This correlates strongly with the increasing influence of other parts of the media company and business outside it. The influence of journalists and politicians is not changing very much, according to the journalists.
- In Sweden, political influence is decreasing (from an already very low level). Other parts of the media company are increasing their influence, and this correlates strongly with the increasing influence for advertisers. The influence of journalists and owners is regarded as almost the same.
- In Russia, journalists are increasing their influence, according to a clear majority (57 per cent) of journalists in the survey. This increasing influence correlates positively with the influence of the audience and the owners. The perceived political influence on media content is almost the same, only slightly increasing.

Table 5.14: *If you look back 5–10 years, how would you evaluate the changes in influence on media content in your news organization from the following actors?* Give your answer on a scale 1–5 where 5 is "increased much"*

	Poland			Russia			Sweden		
	M	SD	N	M	SD	N	M	SD	N
Journalists	2.91	1.07	396	3.57	1.05	398	2.95	0.95	348
Other groups in the media company	3.43	0.87	336	3.17	0.97	391	3.36	0.83	318
Media owners	3.68	0.92	373	3.53	1.11	391	3.08	1.02	309
Politicians	3.14	0.98	364	3.27	1.92	393	2.62	0.78	309
Audiences	3.60	0.96	390	3.53	1.01	407	3.56	0.80	340
Advertisers	3.83	0.93	375	3.01	1.16	381	2.97	1.06	305
Private business	3.41	0.90	341	2.55	1.13	341	2.93	0.85	299

** Only those with 5–10 years of experience could answer the question. Therefore 100–150 answers are missing in each question. (M=mean, SD=standard deviation, N=number of answers)*

Source: The "Journalism in Change" Project Survey, 2012.

Influence on media content is a question of power. Usually, when changing power relationships are analyzed, one group increases its power at the expense of other groups. This is also the case in the answers from Sweden and Poland – there is a negative correlation between influence for journalists and influence for owners, other parts of the media company and advertisers. When journalists in these countries feel that the power of owners and markets is increasing, they perceive their own influence on media content as decreasing (see Table 5.15).

This is not the case in Russia. There is a quite strong correlation between increasing influence for journalists and for the rest of the company, as well as for owners, audiences and private business. The only exception is political influence, which is considered to be slightly increasing, but is not correlated to the influence of journalists. One explanation for this can be that Russian journalists feel their influence is growing stronger when the whole media company is strong – in relation to the political system.

Both in Russia and in Poland there is a clear positive correlation between the influence of the audience and journalists. This indicates that journalists in Russia and Poland in some way see audience (and market?) influence as a kind of ally to their own influence – in relation to the old political influence. In the Swedish media system, journalists feel all other influence has a negative neutral impact on their own influence.

Table 5.15: Correlation between changes in influence on content for journalists over the last 5–10 years and the influence of other groups (Pearson correlation, only significant correlations are given)

Influence of…	Influence of journalists on content in…		
	Poland	Russia	Sweden
Other groups in the media company	-0.144** N=334	0.393** N=381	-0.299** N=314
Media owners	-0.149** N=367	0.168** N=378	-0.477* N=305
Politicians	–	–	-0.136* N=306
Audiences	0.306** N=379	0.241** N=392	–
Advertisers	-0.203** N=368	0.106* N=368	-0.223** N=297
Private business	-0.165** N=339	0.218** N=332	-0.246** N=297

** *Correlation is significant at the 0.01 level (2-tailed)*
* *Correlation is significant at the 0.05 level (2-tailed)*
Source: The "Journalism in Change" Project Survey, 2012.

5.4.2 Political pressure on professional autonomy

Questions about political and economic pressure on professional autonomy are important in almost all interviews with journalists. They give a lot of examples of this, but also examples of how it is possible to deal with these pressures.

Political pressure is clearest for journalists in Russia and partly in Poland. In Sweden, no journalist in the interviews mentions any political pressure at work. Only one reporter on a trade union magazine talks about how they relate to the owners, the trade union for state employees; the magazine is independent, but the union has one page where they can decide amongst themselves about the content [SR14]. Other Swedish reporters talk about the flow of press releases and calls from public relations agencies to make them cover certain issues. But the journalists declare they make a very hard selection and decide themselves what to cover.

Poland has had a higher degree of politicization of media in history. One old journalist on a weekly magazine tells about how he in the 1980s had to go through a process of "verification" to continue his work. He was lucky and regarded as politically correct, and was placed on the magazine "Motor" even if he had no car and no interest in technology [PR18]. Political influence has had the clearest influence in state radio and TV. A reporter on a state TV channel for more than 20 years says of course she feels political pressure and she describes the political influence:

> "I have been thrown out twice, but I have come back every time. Of course I feel it (…) In public TV, people for certain positions are appointed depending on the current political conditions. Everyone knows about it and it is talked about, but not in detail, although it would be good if there were detail (…) any person removed from the management of public television usually seeks revenge at the earliest convenience… and every time it gets worse. It is a phenomena that winds up itself, from what I have observed. It is like a political hamster running around producing revenge after revenge. This of course eliminates any kind of independent journalism. [PR5, female, between 35–55 years old, TV]

One freelancer, previously working for state radio, talks about how the local channel gave a lot of attention to one candidate in the elections, but not the other side. There was a clear political trend in the coverage, and because of this politicization she left the channel [PR13].

In the newspapers the relationship to politics is less clear. One old reporter on the well-known magazine "Polityka" says they discuss the political line in the newsroom, but there is no pressure from owners concerning politics. Another reporter on a local newspaper owned by a German company says he does not even know what political direction the owner has. On a tabloid newspaper a reporter talks about having different reporters covering different parties – but this system became too expensive. Still, there is a political dimension in journalism, and some of the Polish journalists in the interviews talk about the disadvantages of political bias. They don't want to be politically connected anymore [PR11].

In Russia, the political pressure on autonomy is clear in most of the interviews. Some of them are quoted earlier in this chapter. One young editor on a national popular weekly describes it like this:

> "Our editors' office has a certain political tone and accordingly a certain political climate. Even though I work in the neutral social area, I cannot promote liberal ideas due to my personal beliefs and general course of the newspaper. Naturally, none of us attacks the ruling party and the powers that be." [RR12, female, 28 years old, weekly magazine, head of department]

An old political observer on a local business paper compares today's relationship with politics with the 1990s when he became a journalist; "there was true freedom of the press, like what thrives in the west today: "Nowadays there is the power vertical on which local and central media depends" [RR5]. The "vertical" mentioned by this journalist is the political power coming down from the Kremlin, dominating all the state and social structures. This is visible in daily work, according to this journalist:

> "In general, young journalists find themselves within strict administrative boundaries and have to follow the political instructions of their superiors, make highlights as they are required to. Information sources are often not chosen but thrust upon them." [RR5, male, 50 years old, local business medium, political observer]

Some journalists working on newspapers connected to the oil industry describe their work as "corporate journalists" – "there is no talk about independence in the company's media… the company's reputation comes above all else " [PR10]. They talk about self- censorship and distributing information prepared by the company. But there are also other journalists in the interviews describing new possibilities – one young sports reporter on federal radio talks about how he, together with colleagues, work to start a new online music radio. He has big expectations:

> "I hope that in 20 years journalists will work on making news and not filling the airwaves with superficial news about government officials. Journalists' attitude to journalism should also change as well as the audience's attitude to journalism. We will no longer be thought of as disseminators of lies, people will trust us." [PR4, male, between 35–55 years old, public radio, weekly magazine]

Nearly all Russian journalists in the interviews express some kind of political pressure, either directly in their work or from development in society in general.

One editor at the state news agency RIA says "we are now living in the era of reaction, it is not over yet: the regime is going to become stricter, for the activity of civil society is low" [RR6]. One year after the interview, in December 2013, his news agency was shut by a decree from President Putin.

At the same time, other journalists give a more optimistic prognosis in their interviews. A reporter on an elite national weekly magazine says they permit political criticism and present conflicts between different political positions:

> "A publication's reputation costs a lot. As of now, our magazine is self-sustaining and we are proud of that. The policy of this magazine as modern media, is to strive for objectivity. For example, we can critique the Ministry of Education and the next day ask its spokesman to comment. "[RR15, male, between 35–55 years old, weekly magazine, head of the national department]

The interviews with Russian journalists give a picture full of contradictions concerning political pressure. Everybody talks about it, but the attitude differs; some accept it as a reality and try to adapt, some try to find ways to handle it and find ways to create some space for professional autonomy.

5.4.3 Sponsored journalism or separation?

The commercial pressure on professional autonomy can be both direct and indirect. Pressure from advertisers, from sponsors and from owners to improve financial results can directly limit autonomy in daily work. But influence can also be indirect – higher demands on production give journalists less freedom to manage their own time. Earlier in this chapter, Swedish journalists give examples of this.

Direct pressure takes different forms in the three countries. Common for different kinds of direct pressure is the eroding border between journalism and commercial parts of the media companies (marketing and advertising departments). Journalists in the interviews emphasize the importance of this separation – it is a question of credibility. A young Polish radio reporter says that his work is not influenced at all by the advertising department; "If it were influenced it would mean that we create news through the prism of money" [PR1]. Swedish reporters give the same picture; "I have no contacts with them, there are still waterproof doors between our departments" [SR8].

At the same time, most journalists are aware of economic dependency. Competition in media markets is high, and to maintain economic results, it is important also for journalism. The influence takes different forms in the three countries:

In Sweden, supplements are started in the newspapers in areas where it is possible to get advertisements, for example on housing, cars and gardens. A reporter on a regional newspaper gives this story:

> "Money is the motive to do things. For example: the readers want positive news, therefore we started a new positive supplement called "Lovely Friday" with a lot of stories. The purpose was to sell ads, but after they decided that the ad-department did not interfere in our work... nowadays the purpose is that the owners shall earn money, once upon a time it was to make good journalism with the money."[SR 6, male, 45 years old, newspaper, reporter]

A reporter on a weekly magazine for women says that the ad-department want to interfere with the content, the advertisers demand more to buy the ads; "it is more difficult for the magazine to resist" [SR10]. Another reporter on a daily newspaper talks about discussions with the marketing department about her business supplement, but also that they never tell her what to write about [SR4].

Another reporter on a tabloid talks about a growing feature in media outlets – advertisements looking like journalism. On the website of his tabloid "Aftonbladet" (the biggest media in Sweden), there are often ads resembling articles; "those making ads write there. It is very precisely called an "advertisement", but visually it is close to an article" [SR19].

In Poland many journalists in the interviews talk about sponsored journalism and conflicts with advertisers and ad-departments in the media. One reporter on a local newspaper says that sometimes the marketing department has agreements about collaboration with advertisers and their events, and in those cases they have to write about it [PR10]. A reporter on a weekly magazine says he suspects that he is writing a sponsored article, but he does not know it for sure, because sometimes sponsorship is hidden; "the idea of total independence is total fiction, no newsroom, so the source is 100 percent independent", he says [PR18]. Another reporter in commercial TV describes a case when the TV channel collaborated with a ski-resort to send news from there:

> "I would like this tendency to reverse. I wish there would be more formal contacts, more distance, less dependency on money and marketing. I would like marketing to disappear (...) though not because then my television would disappear. And you see – I am already a hostage of marketing". [PR14, female, over 35 years old, TV]

In Russia, journalists also talk about sponsored articles. One political observer on a local business paper talks about how he writes biased articles for customers under a pseudonym and gets paid beside the regular salary for this [RR5]. There are contrasting opinions about this kind of work among Russian journalists. As an editor on a news agency says, "ordered stuff and latent advertising is, in my opinion, one of the generic diseases of Russian media" [RR6]. But another journalist, a middle aged reporter on a federal newspaper has no problem with this:

> "Product promotion involves good earnings. I do not refuse the use of press releases as well as proposals to write an article from advertising executives, for it brings good incomes (...)Some people believe that working with advertising information is unacceptable in every way. I do not agree with this." [RR13, female, 57 years old, federal newspaper, political journalist]

The commercial pressure on autonomy is clear in all three countries. But the interviews give diverging pictures on the changes; in Sweden journalists still want to keep the old "wall" between journalism and the commercial parts of the media company, but at the same time they experience increasing commercial pressure. In Poland and Russia, some of the journalists see the opposite development, towards a clearer separation. Many of the journalists in Poland and Russia perceive the mixture of journalism, advertisements and sponsoring as bad for credibility. A young Polish radio reporter is satisfied that advertisements, marketing and journalism have been separated in different departments, "before a journalist had to do everything herself" [PR11]. An editor on a national weekly magazine in Russia describes the same development:

> "Regarding journalism as a whole, it has become cleaner. There are decent publications, the notion of advertising and article are clearly separated. It is currently a priority for Russian Reporter and other publications. In my opinion in the 1990's almost all articles were bought and sold, there was no such cleanliness." [RR15, male, between 35–55 years old, national weekly magazine, head of the department]

Generally, the level of perceived autonomy in daily work is quite equal in the three countries according to the journalists; a little bit lower in Russia and a little bit higher in Poland. But journalists experience different kinds of pressure in the newsroom in selection of subjects, and this also has a clear correlation to the autonomy in daily work; political pressure has the strongest influence on autonomy, and this is clearly strongest in Russia. But there is also pressure from advertisers and from commercial parts of the media company, and this is strong in Poland and partly also Sweden.

The influence on content is changing, and the patterns differ between journalists in the three countries. Russian journalists feel they can increase their influence together with owners and all other groups except politicians. In Sweden and Poland, there is a clear negative correlation – when owners and the rest of the company increase their influence on content, the influence of journalists' decreases.

5.5 Conclusions and discussion

Journalists in Poland, Russia and Sweden have some kind of autonomy in their daily work, at least according to answers from 1,500 journalists in a survey (2012). Nearly half say they almost always are able to get a subject covered and 28–29 per cent say they have almost complete freedom to select stories and to decide which aspects should be emphasized. But these figures are not equally distributed among all journalists – some say they have more influence in daily decisions and some say less. The purpose of this chapter is to analyze the patterns in these differences, and to analyze to what extent media development is changing these patterns.

The perceived autonomy in daily work has been analyzed on three levels – personal, organizational and societal. This analysis follows earlier research (Reich and Hanitzsch, 2013), and takes it a step further, including new variables connected to media development as downsizing of newsrooms, temporary jobs and multiskilling (RQ1):

Personal level – professional experience, age and professional identity are important indicators of perceived autonomy. More experience and a more important position in the newsroom gives higher autonomy. There is a clear difference between genders; male journalists are reporting higher levels of professional autonomy in all three countries. There are also clear differences when it comes to forms of employment – a steady job gives a higher degree of perceived autonomy than a temporary job.

There is a clear correlation between professional identity and the degree of autonomy in daily work. The interviews show that a certain degree of autonomy is important for all journalists. They are all aware of givens limits for this autonomy (time, format and different kinds of pressures), but still try to find ways to handle this. Those journalists saying they are proud to be a journalist and planning to stay in the profession, also report a higher degree of autonomy in daily work.

Organizational level – different business models give small differences and no clear pattern in perceived autonomy. Only one group shows a clearly diverging pattern: state TV/radio journalists in Russia reporting much lower figures than other journalists. Other factors connected to the organization are generally more important – the degree of freedom to manage time during work and if there has been downsizing of the staff in the last five years. Shrinking staff in the newsroom and less freedom to manage time gives less autonomy in daily work. One important factor in work is positively related to perceived autonomy; those who have more time for research and verification also report higher autonomy in daily work.

New practices in work like multiskilling are often described as a way to increase production and cause negative reactions among journalists (Quandt and Singer, 2009:135). In the survey the figures point also in a direction towards a more positive attitude; journalists with a positive attitude from their own experience in multiskilling also report a larger perceived autonomy in daily work. Journalists working in organizations where they are expected to be multiskilled, have a higher degree of influence to get subjects covered. They can select their own stories and angles, and they agree multiskilling gives a higher quality in journalism. Also when it comes to interactivity, there is a clear positive correlation between positive attitudes towards influence from the audience and daily autonomy for the journalists. It seems as if journalists can take the audience as an ally in striving for professional autonomy.

Societal level – the three countries in the survey are very different when it comes to media system. But still, there are only small differences in the level of perceived autonomy for journalists in daily work. In the survey, there is no direct connection between the opinions on press freedom and autonomy in daily work.

Looking closer at the results, there are important differences. The political influence on content is reported much higher in Russia and partly Poland than in Sweden, and journalists experiencing a strong political influence also report less freedom in daily decisions. This political pressure has also been found in earlier surveys comparing Russian and Swedish journalists (Nygren and Degtereva, 2011:10). Political pressure is the most important limitation to professional autonomy in daily work, confirming results from Reich and Hanitzsch (2013:149). But there are also correlations in pressure from advertisers and business interests, and less autonomy in daily work.

Nearly half of the Russian journalists in the survey say press freedom in the country has decreased in the last 5–10 years (see figures in Chapter 7.2.4). In spite of this, a majority of journalists declare that they have increased their influence on content together with the owners and other parts of the media company. The interviews give many examples of political pressure in daily work, but some of the journalists also give examples of increasing autonomy both in relation to political and commercial pressure. The Russian result is a paradox; both decreasing press freedom and increasing influence for journalists. One explanation is that the results concern two different levels – there can be a growing professional autonomy on a basic level in daily work within Russian media companies (not all, especially not state radio and TV), and at the same time a stronger pressure from the Kremlin. Another explanation found in the interviews is self-censorship. With a strong awareness of the political limits, Russian journalists are still trying to find a space for professional autonomy in the work, in daily decisions about selecting stories and emphasis.

How is media development influencing perceived professional autonomy in daily work (RQ2)? Fears are often expressed that current trends make the position of journalists weaker. Waisbord describes social changes in society and de-professionalization when he discusses professional autonomy (Waisbord 2013:47), and other researchers discuss how transformations in the news industry and business interests undermine professional autonomy (Witschge and Nygren, 2009:55; Örnebring, 2010a:64; Lee-Wright, 2012:38). Other researchers describe a transformation of the professional role from the old – fashioned journalist to a multiskilled media worker with a portfolio work life going in and out of projects. These new media workers gain a new kind of autonomy and security through their competence and networking abilities (Deuze, 2007).

The survey show results in both directions:

- Long experience and strong professional identity are positive indicators for high degree of autonomy. If journalism changes to a profession "passing through", this can make these indicators weaker. In Russia, nearly 70 per cent of the journalists in the survey are under 35 years of age.
- Temporary employed journalists in the newsrooms report clearly lower levels of perceived autonomy, and if the share of temporary employed workers increases, it can have effects on autonomy. On the other hand, freelancers report equal or higher autonomy compared with those with a steady job.
- Downsizing of the newsrooms and less freedom to manage time at work show a clear correlation with lower perceived autonomy.
- Those who describe themselves as multiskilled and having these demands in the newsroom, report higher levels of autonomy in daily work. They feel more freedom to choose subjects and decide angles.
- Those journalists with a positive attitude towards interactivity also have a larger degree of autonomy in daily work. Journalists seem to consider audience influence as an ally in the influence on content.

Many results in the survey indicate a weaker journalistic profession, for example a low share being members in professional organizations and unclear borders towards other areas in media like public relations and communication (see Chapter 3.5 and 6.3) This concerns especially Poland and Russia, and confirms the picture of a society where professions are becoming weaker (Freidson, 2001:185). In all three countries, journalists report a strong market logic in relation to professional logic, but in both Poland and Russia interviews also emphasize the need for separation between journalism and commercial parts of media companies. At the individual level, the tendencies can also be in the other direction; multiskilling and interactivity can increase the degree of autonomy in daily work for strong individuals.

Poland, Russia and Sweden have media systems very different in many aspects – history, patterns in media use and degree of political parallelism. Is the development the same in all three countries when it comes to autonomy in daily work (RQ3)?

Despite the big differences in media systems, the survey shows quite similar figures on perceived autonomy in daily work in the three countries. It seems like a certain degree of autonomy in daily work is part of the professional identity in all three countries, in different media systems. In the interviews, journalists clearly show an awareness of the limits of this autonomy; time and demands in the production, formats and profile, commercial and political pressure. But within these limits, journalists try to find a space for their own initiative. Those creating this space also report a larger professional pride and plan to stay in the profession.

The differences between the countries are the kind of limits and how narrow these limits are. In Sweden, the limits often concern time and freedom to manage time, but also pressure from the commercial parts of the media company to produce content suitable for advertisements. In Poland, some journalists still talk about some political pressure, but the most important limits seem to be related to commercial demands; both pressure from advertisers and fewer resources in the newsrooms. In Russia, political pressure is clear in both newsrooms and society and the limits for professional autonomy seem narrower in relation to political pressure. At the same time, there are signs of increasing influence of journalists and in some interviews also a greater awareness concerning the borders around the profession. In Russia journalists seem to consider commercial influence to be an ally in relation to political influence.

This means there are still important differences between media systems also when it comes to professional autonomy. But there are also signs of change, starting from different points of departure. This change goes along the track of professionalization in Russia, with journalists struggling for professional autonomy. In Sweden and Poland there are signs of a reverse move on the same track, reducing the strength of important indicators for professional autonomy.

6. Ideals and values of modern journalists: the search for balance

Maria Anikina

Quite often the discussion about ideals and values in the field of communication is described through opposing one professional standard to another, through comparing contradictory trends. As Maras writes:

> "Standards of journalism do not stand above history, and what is good or bad practice needs to be reinvented, not just through robust and rigorous discussion in the profession, but through dialogue with new theoretical perspectives and cultural change." (Maras, 2009:207)

Modern scholarly literature describes contradictory trends existing in contemporary journalism (e.g. market orientation and social service of media in different countries and regions of the world, liberalization and regulation strengthening typical for the precise segment of media system etc.). However, all the trends are indivisibly linked to historic conditions, a wider national and international context in which values and standards of journalistic profession are formed and realized.

The following chapter includes analyses of professional choice and its factors, covers the issue of professional ethics seen by practitioners in three countries and represents the overview of value orientations typical for journalists in Poland, Russia and Sweden.

6.1 Professional choice

Professional choice is one of the most serious decisions people make in their life, and the factors influencing this process are typical for a concrete media system and reflect peculiarities of certain social groups (see Table 6.1)

Table 6.1: Factors influencing job choice

Factors	Poland			Russia			Sweden		
	N	SD	M	N	SD	M	N	SD	M
Salary	474	1.090	4.1	499	.818	4.1	481	.867	3.7
Benefits	441	1.331	3.2	480	1.203	2.9	471	1.025	3.1
Editorial policy	454	1.153	3.9	497	.955	4.1	480	.909	4.1
Job security	450	1.205	3.9	484	1.188	3.3	475	1.060	3.9
Chance to develop a specialty	476	.804	4.5	491	.920	4.4	479	.950	3.9
Atmosphere in the workplace	469	.839	4.4	493	.850	4.3	477	.694	4.4
To serve the public interest	476	.995	4.2	495	1.717	3.9	477	.971	3.8
Freedom in the job	475	.797	4.4	493	.911	4.1	483	.789	4.2
Chance for a better career	468	.963	4.2	488	1.144	3.8	479	1.001	3.5
Influence society	457	1.088	3.8	445	1.108	3.5	476	.984	3.7

Source: *How important are a number of things in choosing jobs for you?* Answers on a scale 1–5 where 5 is very important. The "Journalism in Change" Project Survey, 2012.

The factors evaluated highly by respondents in all three countries are the atmosphere in the workplace and the amount of freedom at work. The latter factor will be presented further from different angles. The third important leading condition is the chance to develop a specialty. It seems that all three mentioned factors could create the perfect combination for successful professional activity and development for journalists being put into a favourable context.

Further analysis of the data lets us see national peculiarities. Russian journalists stress salary as meaningful for job choice (4.1) and editorial policy of media company (4.1) which differentiates the Russian case from the other two, and leads to the debates on the social status and the autonomy in different national contexts.

Higher rates of such factors as chance to develop a specialty given by Polish and Russian (means are 4.3 in both countries) in comparison with moderate Swedish evaluation (3.7) tell us about the peculiarities also in terms of stability and instability of media systems in the three countries. Transforming a journalistic culture developed on the basis of social, political and economic changes which happened in recent decades "produces" professionals who are striving for self-improvement and professional development in a more active way than representatives of more stable societies do. It is interesting that socially oriented factors are not as attractive for modern journalists – the opportunity to influence society (3.7) and serving the public interest (3.9) are located in the second group of factors.

An explanatory factor analysis of determinants influencing job choice detects three leading factors among 11 items (see Table 6.2).

Table 6.2: Factors of job choice (Factor analysis, correlations within each dimension)

	Social development	Working conditions	Creative growth and development
to serve the public interest	.838		.217
freedom in the work	.544		.219
influence society	.501		.450
atmosphere in the work place	.382	.367	
influence society	.308		
benefits		.635	
job security		.635	
salary		.587	
editorial policy		.383	
chance for a better career			.855
chance to develop a specialty	.287	.200	.552

Source: *How important are a number of things in choosing jobs for you?* The "Journalism in Change" Project Survey, 2012.

The factor "social development" refers to the inspiration of journalists to serve the public interest, to work with public opinion and to influence somehow the public and society that could realize itself in an independent and fruitful atmosphere when journalists have enough freedom in their work. The factor "working conditions" illustrates a vital, practical dimension of journalism, which includes such important aspects of professional activity as salary, security and benefits in a complex with editorial policy and general working atmosphere. The factor "creative growth and development" shapes mainly creativity and growth potential of the profession. To a certain extent it returns to the creative nature of journalism and traces back to the literary tradition in analyzing journalism.

As Bak writes:

> "Literary reportage and literary journalism are much the same when they both emphasize narrative and descriptive modalities and eschew discursive polemic." (Bak, 2011:8)

In changing conditions of the modern communication process the actuality of literary journalism is discussed, the necessity of narrative modality is not unambiguously evident. But a further description of professional values (see Chapter 6.4) carries on this dimension.

It is interesting that the difference between the three countries is not so evident when we compare the defined and above described factors in Poland, Russia and Sweden (see Table 6.3).

Table 6.3: Dimensions in the choice of workplace (Means on the three most important factors in each dimension)

	Poland	Russia	Sweden
Factor 1 – social development	3.86	3.70	3.76
Factor 2 – working conditions	3.43	3.36	3.40
Factor 3 – creative growth and development	3.90	3.83	3.50

Source: The "Journalism in Change" Project Survey, 2012.

As seen from Table 6.3 the least important integrated factor for journalists in all the countries is related to working conditions. We could suggest that journalism still remains a socially oriented type of activity and people involved in it in different countries pay a bit more attention to the social significance of the profession than to practical matters of their job (such as job security, benefits etc.) despite instability, difficulties and uncertainty journalists face in transitional countries. The salary as a part of working conditions integrated factor is still a problematic issue for respondents and its meaning will further be described separately.. The important finding here is the role of social factors influencing job choice in Poland, Russia and Sweden – the means are almost similar and the highest if we compare the three strongest dimensions obtained during factor analysis of the survey data. The assumption about strong meaning of social orientation that journalists demonstrate is indirectly confirmed in the Swedish case – the dimension comparison shows that Swedes are less inclined to seek creative development and personal success than Polish or Russian journalists.

Thus the analysis clarifies more similarities than differences in the professional consciousness of the respondents in the three countries with different social, political and historical conditions and this could be the reason to discuss supra-national characteristics of the profession realized in different national contexts.

6.2 Professional identification

The level of popularity of journalism as a profession in various societies differs and it leads to the decline or to the increase of trust in journalists and media in different regions. Nevertheless, the conducted study points out that representatives of Polish, Russian and Swedish media feel themselves satisfied with professional choice and leading activities (see Table 6.4).

Table 6.4: Professional identification: agree and fully agree (4–5 on a scale 1–5)

Statement	Poland		Russia		Sweden	
	N	%	N	%	N	%
I am proud to call myself a journalist	477	67.8	489	66	494	78
The main reason for the work as journalist is the salary	492	68.2	480	58.8	495	2.4
My work means more for me than family and leisure time	485	25	492	13.6	494	3.6
My profession is the mission of my life	484	15.8	486	21	494	18
To work as a journalist is a step in my career to other areas	460	44.2	477	32.2	480	5.8

Source: *Here are some statements about your relation to your profession. Do you agree or disagree with the statements?* The "Journalism in Change" Project Survey, 2012.

The tools of descriptive statistics help to find out that the most shared opinion by journalists in all three countries idea is the pride of being a professional in journalism (4.0). A more detailed comparison of the three countries' data does not clarify serious national specifics in this respect.

Data evaluation gives a reason to suppose that Polish journalists are more enthusiastic about professional involvement than Russian and Swedish respondents. This assumption could be indirectly confirmed by the World Values Survey, which shows that Polish citizens state that work 'rather' and 'very important' more often than Russian and even Swedish citizens in general. Thus it is possible to consider social context in Poland a bit more advantageous for the formation of setting and motivations in journalism as a labor activity.

At the same time the "Journalism in Change" project presents idealistic and realistic views on journalism. They can be found in opinions expressed by Russian media representatives during in-depth interviews. Respondents from older age groups demonstrate a balanced and objective position:

> "Today there are few real journalists who perceive writing as a profession. Journalism is a special view of the world, it is responsibility, it's about broadening horizons, the ability to debate matters in mind, about feelings and being responsible for what you tell. Reading texts, I often see that this is a pure provocation, and people who wrote it are not aware of the consequences that may arise from such material. About 70 per cent of Russian journalists, in my opinion, refuse to accept the consequences of what they write and do not feel they are responsible". [RR8, female, 53 years old, editor, Moscow city paper]

Speaking about generational (age) differences in professional identification it is possible to mention several interesting findings (see Table 6.5).

Table 6.5: Evaluation of the statement "The main reason for the work as a journalist is the salary" in different age groups in three studied countries: agree and fully agree (4–5 on a scale 1–5)

Age	Poland		Russia		Sweden	
	N	%	N	%	N	%
Under 35 years	195	66.7	347	57.3	126	1.6
36–50 years	181	65.7	117	58.1	170	2.4
Over 51 years	84	71.4	36	75.0	178	2.8

Source: *Here are some statements about your relationship to your profession. Do you agree or disagree with the statements?* The "Journalism in Change" Project Survey, 2012.

First of all, it is worth noting the radical difference between Poland and Russia on the one hand and Sweden on the other in perception of the financial aspects of professional activity. It is evident in all age groups as presented in Table 6.5. Probably this is an argument to prove a strong social orientation and – on the contrary – weak striving for personal benefits typical for Swedish journalists. At the same time, this picture could also illustrate a wider social and economic process in Central and Eastern Europe, which leads to the shift of motivations to enter the profession. Economic uncertainty forces people to look for well-paid jobs. Another finding is that Swedish respondents of different ages are more homogeneous in their attitudes within the national sample than Polish and Russian ones. Generational specifics in the latter cases correlate with general social trends. Representatives of the elder generation of respondents in

Poland and Russia agree that the main reason for work as journalists for them is salary even more often than younger ones. In the Russian case the situation is brighter – as data show the generational difference here is even bigger, after 50, people are not quite so well-off and often do not feel themselves secure. The above described could be related to bigger risks people face and to the short horizon for future planning described by Gudkov when diverse social groups do not make long-term plans and people try to provide their living in a concrete period of their life.

Thus it is possible to make the distinction between the studied countries looking at national and age groups. Age difference stays evident and at the same time national specifics diminish when respondents describe journalism as the mission of their lives (see Table 6.6).

Table 6.6: Evaluation of the statement "My profession is the mission of my life" in different age groups in three studied countries: agree and fully agree (4–5 on a scale 1–5)

Age	Poland		Russia		Sweden	
	N	%	N	%	N	%
Under 35 years	195	19.0	347	22.5	126	22.2
36–50 years	181	14.9	117	20.5	170	21.8
Over 51 years	84	13.1	36	8.3	178	12.9

Source: *Here are some statements about your relationship to your profession. Do you agree or disagree with the statements?* The "Journalism in Change" Project Survey, 2012.

Overall estimations are quite moderate. This considers journalists to be realistic people without excessive expectations towards the profession and personal involvement in it. On the contrary, to the above presented judgment, the older group disagrees with the proposed statement more often than youngsters do – the acceptance of the statement is decreasing with the age of respondent. The lowest level of acceptance was found in Russia, which could be the consequence of the social and political transformations in recent decades. A fixed general trend could be explained probably by the influence of professional experience and some tiredness from activity in changing and sometimes harsh conditions typical for older professionals. Moreover, it is possible to pay attention to the serious difference between age groups. This difference is strongest in Russia, moderate in Sweden, and almost invisible in Poland (a fact which suggests that this last country has gone through a more balanced and successful development).

It seems meaningful that – even taking into account a certain degree of inspiration towards the profession and general deep involvement into this activity – respondents could not absolutely agree that journalism is the mission of their lives. The attitude towards the profession becomes in certain cases pragmatic – for example the Russians surveyed quite often see journalism as one of the rungs on the career ladder. This position will be further proven.

The attitude towards the profession and the level of satisfaction influences the data reflecting future plans of the respondents. The less positive picture is seen in the Russian segment of the sample – almost one fourth of those surveyed plan to change the field of labour activity – Polish (13 per cent) and Swedish (9.5 per cent) figures are lower (see Table 6.7).

Table 6.7: Personal future as a professional (in per cent)

Country	N	Same company / other media company	Other field than journalism	I will be retired	Do not know / No reply
Poland	490	65.6	13	4.8	16.6
Russia	434	59.4	24.8	2.2	13.6
Sweden	489	63.4	9.6	12.4	14.6

Source: *Where would you like to be working in five years from now?* The "Journalism in Change" Project Survey, 2012.

The presented information is related to the portrait of a journalist in Chapter 4. Also it is possible to note the uncertainty in views of Russian journalists – more than one tenth of respondents could not define their perspectives even roughly. This figure indicates the specifics of social situation in which professional activity is developing. As mentioned earlier the depth of planning horizon (Gudkov, 2014, online), reduced in recent decades and this common trend is reflected in the professional consciousness of journalists.

The majority of those surveyed feel quite satisfied with their position and status – this assumption finds confirmation in the intention to stay in the media sphere demonstrated by almost two thirds of journalists (from 59.4 per cent in Russia to 65.6 per cent in Poland, respectively).

The intention to change the type and field of activity stated by a part of the respondents is linked to the concept of social lift, which is presented in the study in quite a clear way. The difference between cultures is more evident here – one-fourth of Russian journalists, 13 per cent of Polish and only 9.6 per cent of Swedish see themselves working in other fields than journalism. Among the specifications given by Russian respondents there are such activities as teaching, scientific studies, creative issues and quite widely represented politics and PR, which correlate to other ideas of respondents.

Table 6.8: Evaluation of the statement "To work as a journalist is a step in my career to other areas" in different age groups in three studied countries: agree and fully agree (4–5 on a scale 1–5)

Age	Poland		Russia		Sweden	
	N	%	N	%	N	%
Under 35 years	195	52.3	347	37.8	126	10.3
36–50 years	181	38.7	117	23.1	170	5.3
Over 51 years	84	39.3	36	8.3	178	2.8

Source: *Where would you like to be working in five years from now?* The "Journalism in Change" Project Survey, 2012.

Generational analysis shows that pragmatic orientation and attitude towards the profession are more typical for younger journalists, which is natural, because people over 50 normally have already built their career, obtained certain social status and do not have big expectations for future changes in their career. In general during the survey the respondents older than 51 years more often expressed the denial of journalism as a career step, at the same time more than 60 per cent of the younger group of respondents accepted this idea to a certain extent.

National peculiarity could be clarified from the data presented earlier. As previously mentioned, Swedish respondents are not seriously interested in career success entering journalism – the level of acceptance of the proposed statement is relatively low in all three age groups. Again, Polish data point out that the national journalistic community is quite homogeneous, unlike Russia, where figures differ at times in diverse age groups and where a younger group of respondents grew up and obtained professional independency after the social transformations and adapted the new set of values easier and faster than the older generation of representatives did. Here it is possible to see the different trajectories of social, political and economic shifts influencing professional communities, e.g. – the journalistic one.

During in-depth interviews journalists gave general evaluation and several of them describe the future of profession from a values and standards aspect:

> "It is at risk of dissolving in the commercial aspect. Then it will be mere infotainment. The other development option is for journalism to stick to the traditions of the "staunch" media that takes into account the interests of the public, including those of different generations. In 5 years we may be able to employ some successful Western patterns. If we manage to keep working in a certain "alkaline" environment we can assume that Russian print media will yield their position much slower than Western ones. Magazines will be just as popular in 20 years. The Russian audience, in my opinion, will need to read print media." [RR17, Male, 49, editor-in-chief of a web-based media, special correspondent of magazine]

A Swedish respondent pays attention to the distant future and to financial issues and – nevertheless – sees the continuation of the profession:

> "Assange wanted to get crowdfunding, a people funded – journalism, hopefully there will be more of that. Successfully decentralized journalism could really be a great success." [SR 20, male, 40 years old, web editor radio]

6.3 Professional ethics

The problem of common ethical standards is still very acute in the professional community. Scholars representing both sociological and philosophical approaches to journalism studies cite this problem as among the most crucial. For instance, post-colonial theory sees that "the discourse of universality historically served to legitimize the exercise of Western power over the non-Western Other" and stresses that:

> (…) the quest for universals in journalism ethics has problematic historical and political implications." (Meyers, 2010: 79)

The survey data show that generally it is quite important to have common ethical rules among journalists in the media company – the average score of this statement in all three countries is 4.2, which correlates with previous studies. This idea is equally and more strongly supported in Poland and Sweden and less evident in Russia (see Table 6.9). Anyway there is the space for discussion about universalism and relativism as ethical strategies in different national contexts. Here it is worth speaking about the journalistic community in broader frames, because:

> "(…) ethics as a part of culture cannot be understood or evaluated apart from the distinct world or society to which it belongs" (Ladd, 1973:5).

Table 6.9: Opinion of respondents about relation journalism and ethics: agree and fully agree (4–5 on a scale 1–5)

Statements	Poland		Russia		Sweden	
	N	%	N	%	N	%
It is no problem for a journalist to work with PR and information	492	20.4	484	29.8	488	12.8
As a journalist, it is OK to receive gifts from sources as long as it does not influence my professional work.	489	20.8	480	29.2	494	4.0
It is very important to have common ethical rules among the journalists in the media company.	488	91.2	491	59.4	491	90.0

Source: *Here are some statements on journalists, ethics and politics. Do you agree or not?* The "Journalism in Change" Project Survey, 2012.

Comparison between the three countries on the basis of employment types' diversification does not clarify the principal differences – despite the expectation that journalists with fixed contracts more strongly related to concrete editorial boards do not have a more distinct idea of common ethics in comparison with those with temporary contracts or with freelancers.

In evaluations of precise ethical issues more similarities could be found between Poland and Russia. Only 12.8 per cent of Swedish journalists agree with the idea that media professionals could work with PR and information. The figures among Polish and Russian respondents are higher and this could be partly explained by the specifics of higher education institutions organization. Public relations as a sphere of professional activity and a field of education is quite new, and often there are now separate departments offering this type of knowledge and competences to students. It should be added that when speaking about one's personal future, a lot of respondents intended to leave journalism and mentioned PR as an alternative sphere of activity, which suggests that in the consciousness of modern journalists these spheres are very close to each other.

The difference is more evident in estimations of the possibility of a journalist to obtain gifts from information sources – a tiny part of the Swedish surveyed agreed with it while one fifth of Polish journalists and more than one fourth of Russian ones supported this statement. Thus more similarity is seen between Russia and Poland in this question. The suggestion about different levels of corruption in different countries could follow further. But the Global Corruption Barometer delivered by Transparency International in 2010–2013 points out that the level of corruption in the media sphere in both Russia and Poland is close to average (in the 2010/11 report in Poland it was 2.8 and in Russia it was 3.2, media in general had 3.1 out of 5) (*Global Corruption Barometer* 2010/11). There are no measurements in Sweden but it seems reasonable to compare this country with other Scandinavian ones, whose rates are also around the average level. Thus there is no direct correlation between the evaluations of journalistic community representatives obtained during the survey and corruption level estimated by the public.

The lack of strict rules and unified standards of professional behavior resume the question about the concept of universal ethical standards in modern information society:

> "(…) global impact entails global responsibilities (…) To recognize the urgency of global ethics is only the start. To agree that news media have global responsibilities does not mean that we have a clear and defensible list of global duties for journalists around the world. To speak of global media ethics is not to provide a systematic framework of principles, standards, and best practices. Moreover, the idea of global media ethics raises tough theoretical and practical questions." (Ward, 2008:144)

Within the framework of the "Journalism in Change" project the most serious problem is seen in the Russian case, where the dominance of relativism as the basis of perception and reflections on the world of journalism leads to a weakening of internal relations in the professional community. The task to elaborate the set of common ethical rules and standards and follow it was described by Russian journalists during the survey in a moderate way, which is lower than in other studied countries. This figure is the highest in the survey segment dedicated to ethics and politics, but it is far from ideal if we look at it separately. Previous studies of the journalistic community gave similar results – traces of relativism ideology were discovered in the Russian section of the "Worlds of journalism" study that:

> "(…) indicates that impartiality, and the reliability and factualness of information, are valued in different parts of the world." (Hanitzsch, 2010:1)

The opinion about the necessity for common ethical codes and standards is indirectly linked to the integrity of the journalistic community. In the Russian case (as the consequence of a weak striving for common rules) we see a low level of participation in professional unions – less than one fourth of Russian respondents (23 per cent) declared that they were the members of any journalistic union. In Poland this integrity is also quite weak (31 per cent) and only the Swedish surveyed demonstrate professional unity – 86 per cent of journalists named themselves members of journalistic unions (see Chapter 3.5.).

The rejection of a professional journalistic association idea is found in the responses obtained during in-depth interviews. A Polish journalist explains his attitude in the following words:

> "For me journalism means independence and if I ever felt in danger, I know that no trade union, absolutely no trade union would help me, and somewhere, sometime, someone can reproach me for belonging to some unions. No matter if it is Solidarity or any other union." [PR1 male, under 35 years old, radio, reporter, editor, chief]

He also explains why other professionals decide to join professional communities:

> "Because they are afraid they can lose their jobs and they hope that if they belong to such a union it will be more difficult for an employer to fire them. It is a notion like that which is deeply rooted in people's consciousness but is not true." [PR1 male, under 35 years old, radio, reporter, editor, chief]

The described problem seems to be serious not only because of possible weakening of media as a social institution but also because it influences the relations with the audience and – widely speaking – society in general. This could lead to the corruption of social relations and the decrease of trust towards media and journalists. Some recent European surveys point out the loss of trust in the media and journalists tell the truth. The context differs. For instance, the decline in Russian practice is not very evident – 69 per cent of Russian citizens who participated in a national survey still considered media and journalists respectable, 17 per cent do not trust them, 13 per cent do not have a definite answer (*Weekly survey*, 2014). But this does not eliminate the importance of the problem, especially in the context when role spectrum and the structure of the profession are changing. That is why the question of ethical orientations and value systems reflected from both theoretical and empirical sides is still crucial.

6.4 Standards and values

6.4.1 Values in news publishing and information verification

Sometimes obtained information helps to create a stereoscopic image of professional practice which proves the cognitive and reflexive abilities of journalistic community members. Analysing the data, we see that respondents' opinion illustrates diverse classifications and could be an example of several trends in the field of journalism and mass communication.

"– What are the most important values for you as a journalist?
– Giving those that are not visible a voice. I have high ideals.
– Are these values also important for other journalists in your media company? In your country?
– No, I don't experience it like that. Many say it but few do it. They might have other ideals, I do not know what they stand for… maybe they just see the job as a job." [SR 3, female 28 years old, freelance journalist]

This quotation from the interview with a Swedish journalist demonstrates that age does not play the only role in professional standards formation, the reflection on what is happening is also possible in different conditions.

Different approaches exist in modern literature to describe and to analyze values in journalism and journalists' professional duties. In "Normative Theories of the Media" Christians describes four roles of journalism: the monitorial role, the facilitative role, the radical role, and the collaborative role. Nowadays, it is necessary to define the changes in the essence of several professional roles:

> "(…) the contemporary challenges arising in the pursuit of the four roles (…) The monitorial role remains at the core of the task of informing the public and is not essentially changed by new circumstances. (…)The facilitative has been associated largely with the encouragement of deliberative democracy, with encouraging debate and circulation of ideas and information in the public arena. (…) The status of journalism's radical role can be assessed against much the same background. Independent criticism and comment matter more than ever on a wide range of issues that are becoming more complex to assess with any certainty. The collaborative role of journalism remains very much in evidence. (Christians, 2009:237–239)

To a certain extent the realization of journalism's social functions and the fulfilment of responsibilities towards society is related to the quality of journalistic information presented to the audience. Taking into consideration the substantiality of such standards as objectivity, impartiality, factualness of information and others from this row, it becomes important to look at the sphere of information verification.

Table 6.10: Evaluations of different statements concerning information verification: agree and fully agree (4–5 on a scale 1–5)

Statements	Poland		Russia		Sweden	
	N	%	N	%	N	%
In my news organization, news is equally verified in all channels (print, online, radio, TV).	450	55.6	493	67	405	36.8
Incorrect facts are often published because of too little control.	474	29.4	475	24.6	465	25.8
It is better to publish unverified information than to be slower than our competitors.	480	6.8	474	10.4	472	3.2
The audience has lower demands on verification in online news than in our main channel (newspaper, TV or radio).	467	53.8	474	37.8	432	25.8
Online journalism is more of a process, and verification can be done during the process and not before publishing.	465	34.6	465	36.8	439	17.4

Source: *The need to verify and the need to publish news as fast as possible are two basic journalistic values that can be in conflict with each other. Do you agree or not agree with the following statements?* The "Journalism in Change" Project Survey, 2012.

The data presented in Table 6.10 suggest that the news published in different channels and platforms by Swedish media organizations are verified in different ways. This distinguishes the Swedish practice from the Polish and Russian, and this fact will be further proven. Another difference in national professional cultures is in the attitude towards the audience. Polish and Russian respondents believe that the online audience has lower demands on information verification twice that of Swedish ones.

Polish journalists (53.8 per cent) tend to differentiate the online version from the main channel of medium in terms of information verification. They more often than Russian (37.8 per cent) and Swedish (25.8 per cent) think that the Internet audience has a lower demand for information verification. At the same time the positions of Polish and Russian journalists are more similar in reflections on the quality of information in online journalism and on the possibility of fact checking and verification during the process (fewer Swedish journalists agree that this is possible).

The surveyed journalists in all three countries tend to evaluate in a negative way the publication of unchecked information in highly competitive conditions.

This fact stresses the value of proven facts respected by the representatives of three journalistic cultures and keeps some stable standards even in changing times and circumstances (at least on the level of general estimations). Around a quarter of respondents pointed to the lack of control that causes the appearance of incorrect case (Polish journalists more often see it as a problem than Russian and Swedish respondents).

Table 6.11: Evaluations of changes in the last 5–10 years in the area of research and information verification (on a scale 1–5 where 5 is much more, in per cent)

	N	Less (1–2)	The same (3)	More (4–5)
Poland	441	19.5	14.3	66.2
Russia	395	19.9	31.6	48.1
Sweden	380	27.7	46.1	26.3

Source: *During the last 5–10 years, daily journalistic work has changed in many ways. How do you evaluate these changes in the following areas, according to your own experience?* The "Journalism in Change" Project Survey, 2012.

More Polish journalists think that the sphere of research in journalism and information verification changed seriously due to technological and other influences in the last 5–10 years (see Table 6.11).

Even the analysis of information about data and facts verification and about the quality of journalistic content provokes the discussion about the control that exists in daily journalistic work. It is possible to suggest that there are different reasons when unproven and incorrect information appears in the mass media. There could be a set of reasons – not only the lack of control but also meaning and influence of internal or external values and standards important for journalists in three countries.

Among the factors influencing the quality of information transferred by journalists the technological ones appear actively in modern conditions, when "mixed news media" emerge that are interactive and online, where "practitioners use many types of technology to create media content" (Ward, 2013:1). Development of new technologies in news production faces the necessity to ask the journalists whether multiskilling influences the information quality and the level of professionalism.

6.4.2 Professional duties

The conducted survey stresses the national specifics of journalistic views on duties of the profession (see Table 6.12).

Table 6.12: Evaluations of professional duties of journalists

Journalists should...	Poland			Russia			Sweden		
	N	M	SD	N	M	SD	N	M	SD
criticize injustice	489	4.4	.671	492	4.0	.957	491	4.4	.865
tell the truth regardless of the consequences	495	4.3	.671	493	3.3	1.084	494	4.1	.919
ensure that the media business does well	487	3.2	1.071	482	3.2	1.119	485	2.6	1.164
be a neutral reporter	493	4.2	.913	496	4.4	.824	493	4.2	1.040
stand free of special interests	495	4.5	.752	488	4.3	.896	487	4.4	.906
bring forward various opinions	495	4.6	.655	492	4.3	.919	483	4.4	.746
simplify and explain	490	4.0	.885	488	3.5	1.085	491	4.5	.703
influence public opinion	482	3.5	1.004	492	3.5	1.094	485	3.1	1.118
investigate government claims	484	3.1	1.123	486	3.0	1.119	491	4.4	.855
provide entertainment	480	4.2	.845	456	2.8	1.154	494	3.3	1.140
provide information objectively	491	4.6	.648	487	4.3	.913	494	4.6	.633
control public authorities	487	4.4	.754	480	3.5	1.210	495	4.6	.653
represent diverse social groups	481	4.2	.920	486	3.9	1.035	476	3.6	1.274
mobilize people to act	487	4.0	.936	475	3.2	1.190	484	3.0	1.146
educate the public	489	4.3	.828	493	3.9	1.055	491	3.9	.929
stimulate new ideas	485	4.0	.914	483	3.8	1.043	485	3.9	.951

Source: *How important are the following professional duties for a journalist in your country? (Give your opinion on a scale 1–5 where 5 is very important)* The "Journalism in Change" Project Survey, 2012.

The analysis of the epistemological orientations spread among media professionals marks out the neutrality and the objectivity as respected standards of professional work with minor differences between the three studied countries (see Table 6.12). At the same time, attitudes towards the truth in journalists' texts differ in the three countries. Swedish (4.1) and Polish (4.3) respondents evaluate the guidelines for journalists "to tell the truth regardless of the consequences" higher then Russian ones (3.3). On the one hand these data turn back to the debates concerning relativistic and idealistic strategies realized in professional activity of journalists. On the other hand the difference in evaluations is not crucial and the analysis of the current situation in which journalists act should be necessary to understand the reaction of respondents.

Answering the question about the most important values, Swedish journalists offer the following position:

> "We have rules of ethics to relate to. It is important to strive to be the consequence neutral; otherwise you have some kind of agenda. I should not bother myself so much with the consequences of what I write, then I somehow take a position." [SR 8, male, 48 years old, newspaper reporter]

This opinion to a certain extent proves the data of the survey. A further quotation from an interview with a Russian respondent could be one more accent witnessing that generational difference and peculiarities do not play a final role in value and epistemological orientation in modern journalistic communities:

> "I wouldn't divide us into generations, as the understanding of the profession does not depend on one's age. There are quite serious 17-year-olds, who reasonably react to criticism, learn how to write, and we quickly start to get along. As a result, young people become part of our team. We are together, and the age never matters." [RR8, female, 53 years old, editor, Moscow city paper]

As noted before, "journalism has a double role in modern society" (Nygren, 2012:6), it could be described as product and service in a market economy on the one hand and as a part of a political and social system – on the other.

The survey showed that the economic (business) aspect of the profession is not very important in daily journalistic work (especially for Swedish journalists, where the mean is 2.6 out of 5). At the same time, the socially oriented roles of journalists were estimated higher by the respondents in all three countries (the Polish data demonstrate a stronger intention of that country's media professionals to lead socially oriented activity in comparison with two other countries).

Also it is evident that the concept of journalism as a profession and as a segment of creative industries at the same time was confirmed during the survey. The task to formulate new ideas was marked as important by respondents (3.9). The analysis of the factors important for job choice and – especially – the opportunity for personal realization, the chance to develop a specialty highly estimated by respondents (see Chapter 6.2) made in parallel with the analysis of journalists' duties comes back to the discussion of journalism specifics and different approaches to define it, e.g. to the concept of literary journalism which provides:

> "(…) intimacy, subtlety, and artistry we need to understand the times in which we live. The ordinary characters, the immersion reporting, the craft and artistry in the writing, the recognition of complicated problems in representing reality – all this give literary journalism a lasting quality when the interest in momentary details has passed." (Sims, 2011:90)

Further factor analysis will show the complexity of journalistic roles and duties. Looking at evaluations given by respondents it is worth mentioning that the control of authorities' activities, social participation and mobilization were not strongly supported by Russian journalists. Polish journalists strongly agree with the meaningfulness of the above -listed professional duties. It is almost the same situation with educational function and potential of journalism. Russian and Swedish journalists are less active supporting it, and Polish respondents could be seen as their opponent in this aspect. Russian journalists are not ready both to educate the public and to provide entertainment – the first option is in the middle of the list, the second is the last according to the data. The intention to educate society is weaker than it was during previous decades (especially during the Soviet period when this niche was better developed than currently and the list of journalism functions was a bit longer (Grushin and Onikov, 1980:146–168). Today we see a gradual revival of enlightening and educational media but this revival has not seriously changed the professional consciousness of journalists. Yet, it is not as brightly seen in the respondents' evaluations as it could be.

Table 6.13: Professional duties of journalists. Factor analysis, correlations within each dimension

	Activist educator	Neutral reporter	Advocate of justice	Investigator	Entertainer
...mobilize people to act	.75		.21		
...educate the public	.68				
...stimulate new ideas	.62				
...represent diverse social groups	.46	.27			
...influence public opinion	.42				.23
...be a neutral reporter		.66			
...bring forward various opinions	.23	.61			
...stand free of special interests		.60	.30		
...provide information objectively		.58	.29		
...control public authorities		.25	.63	.32	
...tell the truth with no regardless of the consequences			.48		
...provide entertainment	.24		.45		.38
...criticize injustice	.31		.42		
...investigate government claims				.63	
...simplify and explain		.22		.50	
...ensure that the media business does well					.48

Source: *How important are the following professional duties for a journalist in your country?* The "Journalism in Change" Project Survey, 2012.

Factor analysis of a set of journalistic ideals (see Table 6.13) demonstrates two main and three additional groups of factors that one may form after the dimension reduction procedure. The factor "activist educator" is characterized by the necessity to mobilize people to act, to educate the public, to influence public opinion and to represent diverse social groups. The factor "neutral reporter" tends to represent the values of objectivity and impartiality in journalism, to introduce various opinions and to stand free of special interests. The remaining three factors do not have such strong evidence in the data. The factor "advocate of justice" deals with the criticizing of injustice, investigating the powerful and discovering the truth regardless of the circumstances. The factor "investigator" demonstrates professional orientation on investigation and explanatory journalism that is poorly represented in modern media practices in general. The factor "entertainer") is quite weak (represented by two main components) but signifies a modern segment of journalism linked to the notion of journalism as a commercial industry and grounded in the necessity to attract a wider audience in order to survive in financially complicated conditions.

The analysis of the first three detected factors' importance in the three studied countries helps to understand the differences and similarities of professional journalistic cultures from another perspective related to the perception of the main journalist's duties.

Table 6.14: Dimensions in the professional duties for journalists (Means on the four most important factors in each dimension)

	Poland	Russia	Sweden
Factor 1 – activist educator	4.13	3.7	3.6
Factor 2 – neutral reporter	4.48	4.33	4.4
Factor 3 – advocate of justice	4.33	3.4	4.1

Source: *How important are the following professional duties for a journalist in your country? (Give your opinion on a scale 1–5) A journalist should be...* The "Journalism in Change" Project Survey, 2012.

As the data show, the most common thing in attitudes to the duties of a journalist in Poland, Russia and Sweden is being a neutral reporter (see Table 6.14). Let us stress that this is a kind of general perception of respondents close to the ideal. On the level of general estimations there is the space for common professional consciousness which is reduced sometimes by professional practices realized in different contexts corrupting the set of important duties, making journalists to violate the rules and standards and – finally – causing the differences of national professional communities.

The dimension of "activist educator" is more important in Poland but less important in Russia and Sweden. Probably this integrated role is not as demanded in Swedish society or there is no mass audience open to receive this type of information in transforming Russia. Another explanation could be related to the role and status of journalism and mass media in different societies where the role of activist educator is less acute than any other roles in a concrete period of time. The same reasons could lead to a gap between perceptions of the third defined factor. The dimension of justice and watchdog is of secondary importantance in Poland and Sweden, making these journalistic cultures closer to each other but less important in Russia, stressing the specifics of the Russian case for which this watchdog role is historically unfamiliar.

Coming back to the general analysis of roles and journalistic duties in modern society, the proposed statements could be grouped into some clusters. Social or commercial orientation could become more evident, collaborative or a radical role could be seen as dominant, Western professional standards known nowadays by professionals in post-Soviet regions could co-exist with propagandist challenges acute for some media organizations, which is related to the gradual historical development of professional culture of journalists in transformation conditions.

Factor analysis of the data of previous studies helps us to find out some similarities and differences in profession perception. The "World of journalism" project dealing with a wider range of twenty three countries, including Russia in 2007–2011, clarified as the strongest the factor "watchdog" represented by such roles as the watchdog of government and the watchdog of business combined with the task of providing political information. The factor "neutral reporter" is a bit contradictory and tends to advocate social change and to support official policies motivating people. The factor "advocate of justice" is realized in detached observation of reality and influencing public opinion. The factor "investigator" correlates to the above – described "entertainer" represented by the task to attract the widest audience through providing people with interesting information. Thus it is possible to point to some close role perceptions with certain exceptions looking at two sets of survey data gathered during two projects.

Coming back to the "Journalism in Change" project, it is also interesting to note that there is no strong influence of pragmatic orientation on the formation of professional ideals. Cross tabulation analysis of the data does not prove the assumption that financially interested in journalism people, are less inclined to advocate social change or to be neutral reporters. Probably this means that there is not only interdependence between job choice and values of the journalists, and the concept of professional journalistic culture is quite a integrated and complicated phenomenon.

Also it is possible to note that the claims that a new generation of journalists who come to replace professionals with a stable system of positive values could stay unjustified while analyzing obtained research materials such as follows:

> "I think it will get better. Now is the time of people who now are about 30 years old (…) Ten years ago journalism was virtually non-existent, only sponsored materials everywhere…, but now I see that the situation is getting better. This generation, the one I just mentioned, they come with their principles, distribution of money is not their objective, at the same time, they are not wide-eyed idealists, they can see the reality. Money is important to them, but they highly value the profession of journalism, defend the principles, these people make me see journalism as it should be. I believe that in 20 years there will be more people like that, I personally see a positive trend, and I like it." [RR1, male, 35 years old, radio and national newspaper freelancer, online media]

6.5 Journalism and society

The context, social and political conditions for journalistic work are different in the three studied countries. A large part of Swedish journalists estimate the level of press freedom in the country as stable. The Polish segment of the conducted study demonstrates relatively optimistic attitudes and declares that the press freedom level stayed the same and even increases in recent years. The Russian group of respondents describe a mainly negative context – 46.8 per cent of respondents say that the freedom level decreased in the last 5–10 years. These estimations correlate to the data presented by diverse international rankings concerning similar areas. That is why it is possible to conclude that journalistic communities in three countries have different "starting positions" and dispose dissimilar situations to realize the mission towards society and follow professional ideals in professional activity.

Shifts in value system and orientations today is less sharp in post-Soviet countries than in was 10–15 years ago, and that would be probably the time to coordinate the systems, to elaborate common norms and to implement them into practice of diverse journalistic cultures representatives.

In the described situation, as previously mentioned, journalists have unequal potential, which could prevent them from complete fulfilment of their obligation. The reality shows that Swedish media representatives are closer to the ideal model – the level of professional responsibilities' fulfilment, and the mean is 3.5 in comparison with Poland (2.9) and Russia (2.7).

Table 6.15: Fulfilment of journalists' responsibilities towards society

Country	Disagree (1 – 2)		Yes and not (3)		Agree (4 – 5)		Don't know / no answer		M (1–5)
	N	%	N	%	N	%	N	%	
Poland	159	32.6	149	30.6	172	35.3	7	1.4	2.9
Russia	190	38.0	205	41.0	91	18.2	14	2.8	2.7
Sweden	61	12.3	143	28.8	284	57.1	9	1.9	3.5

Source: *Do you think that the journalists in your country fulfil their responsibilities towards society? (Give your answer on a scale 1–5)* The "Journalism in Change" Project Survey, 2012.

Table 6.15 demonstrates the detailed differences in perception of responsibilities fulfilment – few respondents could not give a precise answer (and chose a "don't know" answer), the rest reply according to their feeling and idea of reality. As presented above, Russian journalistic practices are quite far from complete realization of mission and main duties – only less than 20 per cent of Russian respondents consider fulfilment complete, among the Polish surveyed, this figure is a bit higher (one third of respondents), in the Swedish case the figure is the highest – more than half of journalists evaluate fulfilment as complete or almost complete.

Certain evaluations of social functions realization exist in in-depth interviews' materials:

> "Today I have a perfectly clear structure of journalistic work. It is the ambition to complete a social mission that is like the mission of doctors and other representatives of the social sphere. Sadly, I think that many young people go into journalism for the wrong reasons. There is a conflict between the profession and ideas about that profession. To put it bluntly, the first priority for journalists beginning their career is vanity, their primary effort is to become a public figure, and only then to fulfil themselves. That path seems direct to them, but in reality it's false." [RR17, Male, 49 years old, editor-in-chief of a web-based media, special correspondent of magazine]

The cited remark of the respondent's concern about the lag between the real situation and idealistic reflection is reasonable. It has an explanation in scholarly literature and finds confirmation in empirical studies. Again the data show that the reality does not match the idealistic notions. But the fact that journalists think about this gap and could evaluate it seems to be a positive sign of professional consciousness work. It is possible to stress that there is the place of coincidence of *real* and *ideal* in the professional consciousness of modern journalists. Despite the fact that the descriptive portrait of journalists "could significantly differ from the miserable reality" (Anikina, 2012:242–243), the previous study of sixty journalists working with extreme situations and accidents held in the project "Mass Media in Russia and Sweden: a comparative study" clarified that problematic situations and conflict context could make the authors follow higher standards and to respect existing professional rules covering such events." (Anikina, 2012:243).

6.6 Conclusions: ideal notions in the real world of daily journalistic practices

The overview of the presented research data indicate the mismatch of real practice descriptions and ideal concepts of professional activity existing in the professional community in Poland, Russia and Sweden. This mismatch becomes less evident when it comes to extraordinary situations – theoretic notions coincide with strict practical rules and high standards of professional conduct in work of journalists covering emergency cases, catastrophes etc. (as it was the Russian case after the Smolensk tragedy in 2010). This leaves space for professional growth and development. But at the same time such a conclusion poses questions about the specifics of such development and applies for a detailed analysis of journalistic content, which could be a proper realistic addition to the discussion of values and standards of journalists' service.

The notion of national journalistic culture is understood as:

> "(…) a set of ideas and practices by which journalists legitimize their role in society and render their work meaningful." (Hanitzsch, 2007:369)

This notion, in relation to the current (and flexible) national context, helps to define the peculiarities of professional consciousness influenced both by external (such as globalization in different spheres of social life etc.) and internal factors. The conducted study proves the integrity and complexity of journalistic phenomenon.

The existence between the opposing poles typical for transitional professional cultures materializes in a balance between social and pragmatic orientations that is seen in a structure of professional choice motivation, in ethical strategies and in a set of respected professional values.

The analysis of job choice in the "Journalism in Change" project stresses the specifics of a creative profession and continues the discussion about creative class initiated by Florida. Indeed it is possible to see journalism today as a profession belonging to the core of creative class engaged in the activity in which economic function is related to the making of new ideas and new contents, whose social influence on politics and life styles is growing (Florida, 2002), which is seen as attractive both for the people defining their future and choosing the place to work and for those who consume creative content. The data show that the factor linked to professional development in the creative sphere represented by few practical components during the survey appears almost as obvious as the two remaining factors (pragmatic orientations and the orientation towards social service).

An ethical strategies examination during the study discovers inconsistencies between three studied journalistic cultures. Journalists in post-Soviet countries (Poland and Russia) are more inclined to combine activity in mass media with other types and fields of activity (e.g. with public relations). According to the study they also have less strict rules in the work with the sources than Swedish journalists have. The third main finding in an ethical perspective refers to the mostly subjective question of professional community integrity and the objective profile of this community in the three studied countries. It is hard to conclude anything unambiguous about the necessity of common ethical rules for all journalists. But the obtained results again raise the question about the domination of ethical orientation, the choice between generalization and differentiation of ethical standards (the attempt to find something common in actions of people in similar circumstances and define it as common standard or to explain each situation as unique) (Singer, 1961), between idealistic and relativistic perception of ethical rules. Experts speaking about the described mismatch of real practice and ideal concepts propose to move towards co-existence and collaboration in striving to understand and to accept specifics and disagreement of others. In this case common understanding of things and consensus are not as important as mutual coordination and understanding of positions of all the actors. One of the main dangerous consequences here could be seen in the incomplete understanding of professional unity. It is quite evident and it should be named among current problems of modern journalism because it may have serious negative consequences for journalists and society in general.

As Maras writes:

> "The roles of media are very important, as well as being commercial enterprises (or run like them) media outlets are important institutions in our public life." (2013:207)

The "Journalism in Change" project indicates that the role spectrum today spreads from interventionism materialization to market orientation. The decade of the 2010s continues with changes in media system models and with shifts of patterns which until recently, seemed to be quite stable. Social and political transformations were changed and accompanied by shifts in the nature and core of the profession. But, at the same time, it is a bit early to predict the end of journalism as a profession both in the sociological and philosophical aspects, since journalism is a matter of:

> "(…) having a professional attitude and accepting one's professionalism in the marketplace of ideas." (Mayer, 2013:113)

That is to say, since we observe a certain degree of understanding and reflection demonstrated by the representatives of the studies of three journalistic communities.

7. Journalism and politics

Bogusława Dobek-Ostrowska

In this chapter we try to find answers to some questions which are linked with the relationships between journalists and media in three countries around the Baltic Sea:

RQ1: Are journalists interested in politics?
RQ2: What political values do they have?
RQ3: Do they express their political preferences openly?
RQ4: Is politics an alternative for them as a professional activity?
RQ5: Do journalists meet any obstacles in their work from political actors?
RQ6: Do they feel the influence of state, political parties and politicians, how do they note it?
RQ7: How do they perceive the level of press freedom in their countries?
RQ8: Is political parallelism present and how is it manifested by journalists in each country?
RQ9: Are there differences among Poland, Russia and Sweden in the relationship between journalists and political actors? Are there any similarities ? How can we explain these similarities and differences?

We hope that the answers to these questions will help us to verify this hypothesis:

H: *The political system and democratic standards determine journalistic culture in Poland, Russia and Sweden and produce the different models of relationship between journalists (media) and political actors (politics).*

We wonder how the theoretical concepts presented above, as a quality of democracy, press freedom, political parallelism, a hybridization of media system, are reflected in the empirical material collected in 2012 in Poland, Russia and Sweden. In the first part, we analyze the personal political orientation of journalists and their interests in politics and political questions in society. We search for their professional values and ethics in a context of relationship with politics. In the second part, we try to find out what journalists thought about their profession and how they evaluated a state of connection between media and politics in each country.

The analysis, based on the survey of 1,500 journalists (500 in each country) and 60 interviews (20 in each country), help us to verify or falsify the general statements and opinions about journalistic professionalism and relations with the political world. Comparative studies are not easy in general, but it seems to be more difficult in the case of three countries, which present such differing democratic standards (*Democracy Index* – rank 2nd, 44th, 122nd in 2012), a level of press freedom (*Press Freedom Index* – rank 12th, 24th, 142nd in 2012) (see chapter 2.1.3.), journalistic tradition (free media vs politically. controlled media), historical experiences (independent state vs. a long time under partition), and ideological background (communism vs. social/liberal democracy).

7.1 Political preferences of journalists and their personal activity in politics

First, we focus on political preferences of journalists and their values at the individual level. Next, we try to answer the question whether ideological orientation can be a factor, which weakens or consolidates the journalistic standards and professionalism, and how it influences the relationship with the political world.

7.1.1 Political preferences of the respondents

We begin this part of analysis with political preferences of respondents because it is a crucial factor which can help to explain other aspects of this study dedicated to the relationship between journalists and politics.

Social democratic ideology is strongly rooted in Swedish politics, the Social Democratic Party being one of the oldest. It was founded in 1889 and was ruling party many times in the history of country. For this reason, it is not surprising that 50 per cent of all respondents (and 62 per cent who answered this question) indicated "left" and "a bit left" regarding their political orientation. Only 14 per cent of surveyed journalists confessed to right and "a bit right", and 17 per cent to centrist political preferences (see Figure 7.1). It should be mentioned that left ideology has dominated the society for a few centuries, and it has never been disgraced as in the case of the post-Soviet region. It is a characteristic feature of Sweden that differentiates this state from several European old democracies where the equilibrium of political ideologies in societies is observed.

In Russia a balance – among the left (21 per cent), centre (26 per cent) and right (24 per cent) journalistic preferences – is visible. About 30 per cent of journalists have not known what political values they shared. An equal proportion of the left (30 per cent) centre (24 per cent) and right (26 per cent) – oriented respondents was noted among Polish respondents. Additionally, an equal amount of respondents in Sweden and Poland (about 19 per cent) replied: "I choose not to answer", "I do not know", or did not choose any answer at all, that is to say, did not indicate any preferences. It can be surprising that in Sweden – in the state where high democratic standards work well – political pluralism among journalists is lower than in Poland and Russia. One explanation of this result can be that Swedish citizens for most of the 20th century voted for the Social Democratic Party, which dominated the political scene. From this perspective, respondents' political orientation reflects the social structure of the voter's preference. On the other hand, many respondents separate their private political orientation from professional activity and political orientation of the media owners, with whom they do not have relations (e.g. the interviews with SR 12, SR 16, SR 18, SR 19, SR 20). It can be explained also by the "ideology" of being critical towards a political power, typically held by Swedish journalists.

Figure 7.1: Political orientation of journalists (in per cent)

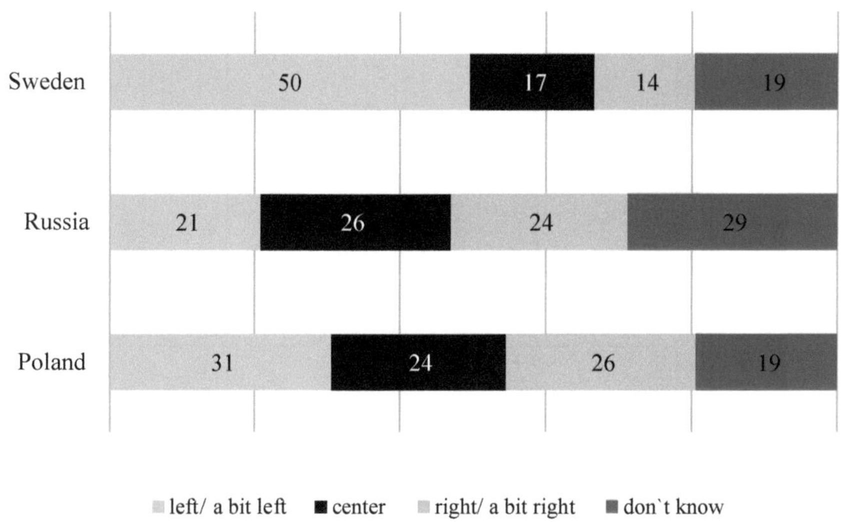

Source: *Where would you place yourself on a political scale from left to right?* (N 500= 100 % in a case of each country). The "Journalism in Change" Project Survey, 2012.

Searching for political preferences of journalists employed in different media sectors one might ask whether there is a similar structure of their orientations as in a general respondents' population? If there are differences, we try to explain it.

Table 7.1: Political orientation of journalists in media sectors

Media sector	Poland					Russia					Sweden				
	N Each sector =100%	Left %	Center %	Right %	No answer %	N Each sector = 100%	Left %	Center %	Right %	No answer %	N Each sector =100%	Left %	Center %	Right %	No answer %
National daily newspapers	51	37.2	23.5	33.4	5.9	110	21.8	30	27.3	20.9	67	44.8	20.9	11.9	22.4
Regional daily newspapers	81	25	35	23	17	63	28.6	22.2	27	22.2	143	42.6	22.4	17.5	17.5
Magazines	127	35.4	24.4	28.3	11.9	153	23.5	20.3	23.5	32.7	94	55.3	11.7	16	17
Public TV stations	46	23.9	10.9	23.9	41.3	43	13.9	44.2	13.9	28	37	35.2	16.2	8.1	40.5
Private TV stations	52	32.7	21.2	26.9	19.2	38	13.1	31.6	23.7	31.6	20	66.7	7.1	9.5	16.7
Public radio stations	64	23.3	26.6	18.6	31.5	7	0	14.3	28.6	57.1	42	66.7	7.1	9.5	16.7
Private radio stations	40	22.5	30	25	22.5	18	5.6	222	44.4	27.8	3	0	0	66.7	33.3
Media online	29	51.7	6.9	24.2	17.2	40	25	15	22.5	37.5	8	37.5	25	12.5	25
News agency	10	20	20	40	20	10	0	30	20	50	21	61.9	12.3	4.8	19
Other	0	0	0	0	0	18	22.2	38.9	11.1	27.8	65	60	13	11	16
Total number of respondents	500					500					500				

Sources: *Where would you place yourself on a political scale from left to right?* (N500 = 100 % in each country). The "Journalism in Change" Project Survey, 2012.

In general, the print press in Poland and Russia is characterized by a balance among left, centre, and right oriented journalists (see Table 7.1). Only in Sweden, a distance between left oriented journalists and others was higher (but less than in the case of the broadcasting media).

Polish journalists employed in the national daily newspapers and magazines declare their political views in a visible way. The percentage of respondents avoiding the answer has been the lowest (about 6 per cent in case of the dailies and 11 per cent in magazines) among all media sectors. Both these sectors are characterized traditionally by an external pluralism. It is manifested by a clear editorial policy, as well as the highest level of politicization. Some national dailies as well as a lot of opinion weeklies address their product to audiences of clearly defined preferences. In this context, publicly declared political views of journalists are their value, besides there are no mechanisms here (like in the case of public broadcasting where more than 31 per cent of respondents in radio and 41 per cent in television do not answer this question) that might cause journalists to hide them. In Russia, journalists employed in the print media were one the largest groups of respondents (65,2 per cent) among all media sectors. There has not been noted a significant difference among left, center and right-oriented journalists. Thus, we can talk about rather balanced political pluralism in the national daily newspapers and magazines. In Sweden, which seems to be evident for this state, the largest group of respondents employed by the national daily newspapers has shared left values (about 45 percent), but it was less than in the total population (50 per cent), which revealed their political orientation in a survey. A right ideology has not been attractive for those journalists (about 12 per cent). Also in the case of magazines, the leftist group has dominated over others (55 per cent).

As opposed to the functioning of national newspapers and magazines in Poland, most respondents employed in the regional dailies have declared centrist views (35 per cent); it is the highest percentage in this category among all sectors. The difference between people with left, center and right preferences has been balanced. This might be explained by the fact that in different country regions (*voivodeships*) there is, in fact, a monopoly of one title and its political involvement might adversely affect the position of the newspaper on the market. In such conditions, one title should be addressed to and reach every reader regardless of their political views (Dobek, Barczyszyn, Baranowski, and Michel, 2013:51). That is why the owners (Polskapresse owned by the German group

Verlagsgrouppe Passau and Media Regionalne by the British company Mecom[1]) have never gotten involved in politics and consistently avoided supporting any political actors.

Thus, it can be assumed that clearly stated political opinions of journalists employed there, contrary to the national dailies, are not in agreement with the expectations of the owners of regional titles. The balance among left, centre, and right oriented journalists in the regional dailies sector in Russia has been visible in contrast to Sweden, where the regional daily newspapers market is very well developed. It was reflected in the number of respondents, who were the most numerous in a survey (about 31 per cent). Evidently, the largest group has at present a preference for the left (about 43 per cent), but it was less than in other sectors.

In the case of Polish broadcasting media, regardless of their ownership status, no clear over-representation of supporters of left, centre or right views has been noticed. The biggest, 6 per cent difference has been noted in the case of private television in which over 33 per cent of respondents have admitted having left-wing/center-left views, in the case of public radio it has been about a 5 per cent difference. Public television (TVP) is an exception, where the proportion of journalists with different views is the same. The highest percentage of journalists who have not declared any political preference or have chosen the answer "I don't know", are employed in public broadcasting – as many as 32 per cent of respondents employed in Polish Radio (PR) and 41 per cent – in TVP. This might be explained by a lack of confidence among employees to their managers. They have had a negative experience – they were "awarded" or "punished" for declaring political views many times in a situation of political power alternation. Perhaps they have not admitted their political orientation for fear of being fired. This lack of response might be a result of a strategy which can help journalists survive in an uncertain situation, and which was noted by one respondent:

> "Especially when it is public radio or television, the current political option becomes so importune, so visible.."[PR13, female, 35–50 years old, freelancer]

In Russia, about 57 per cent of journalists working for state radio stations and 32 percent of journalists working for private TV stations have not wanted to reveal their preferences. More respondents declared center preferences, apart from those employed by private radio stations (more than 44 per cent have declared right orientation). Sweden seems to be quite different on this map. Almost

1 The British company Mecom, the owner of Media Regionalne, left Poland in March 2013.

67 per cent of journalists who have worked for private TV stations and public radio stations revealed a left orientation.

7.1.2 Politics as an alternative

Political orientation is a very sensitive issue in the case of the journalistic profession. A journalist can keep it as his/her private idea, present it publicly and use these views in a professional activity, be ready to work in politics. For this reason, we have asked the respondents about politics as an alternative profession (see Figure 7.2).

Figure 7.2: Politics as an alternative profession (in per cent)

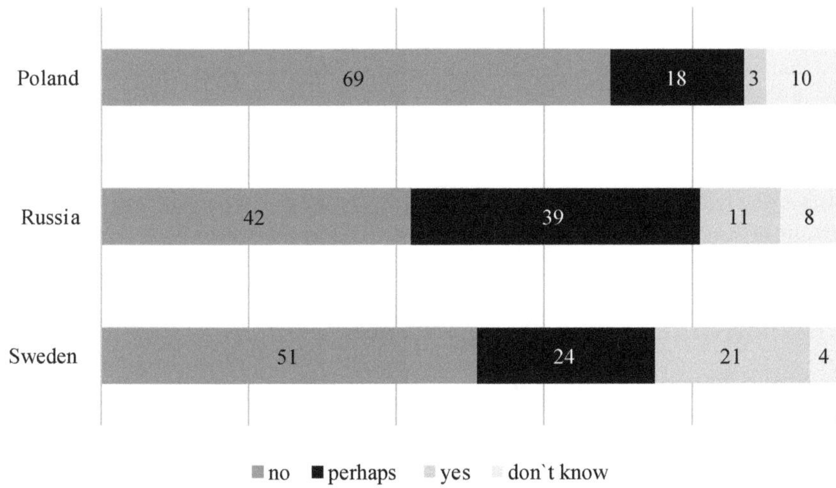

Source: *Is it an alternative for you to be active in politics (voluntary or as employed)?* (N 500= 100 % in the case of each country). The "Journalism in Change" Project Survey, 2012.

The largest group of journalists, who do not see an alternative to be active in politics, are noted in Poland (69 per cent of all respondents but over 76 per cent of those who answered this question). It is a visible difference in comparison with the other two countries. We can explain this by the negative experience of Polish journalists with political actors during the last two decades, where party logic dominated in the coverage of politics in public radio and television stations. The public broadcasting service suffered a lot by strong politicization and instrumentalization at that time. During the presidency of Aleksander Kwaśniewski (1995–2005), the majority of managerial staff confessed a left orientation. A lot of journalists lost their job as a consequence of the presidential elections in 2005 and 2010 and power alternation. Journalists from public radio and TV stations discovered that it would be better not to express their personal political orientations, not to support any party and not be active politically.

In general, Polish journalists like to write and talk about politics and politicians, criticize the political world and comment on political events, but they do not see a place for themselves on the political scene. The results, which we have received in a survey, support this opinion – only 3 per cent of Polish respondents have answered "yes", by contrast to 21 per cent (the largest group among the three countries) in Sweden and 11 per cent in Russia. About 40 per cent of Russian and 24 per cent of Swedish respondents have considered their active role in politics (answers "perhaps"). It is interesting that this same feature of journalists is perceived as something natural and does not raise any negative connotation in Sweden, and it is deeply criticized in Poland. One explanation for this result is, in contrast to Poland, that the overlapping of three spheres of journalists, politics and public relations is limited or does not exist in fact in Sweden. They are nowadays functionally separate, despite a long history of strong press party. In Poland and above all in Russia, those three spheres are more integrated.

7.1.3 Interests in politics and political questions

In general, journalists in the three countries are interested in politics and political problems in society (see Figure 7.3). The biggest group is visible in Sweden (more than 81 per cent) and Poland (74 per cent). In both countries other categories ("not interested" and "middle") take 18 per cent in Sweden and 23 per cent in Poland. A different structure might be observed in Russia. About 55 per cent of respondents have been interested, but a quite high group of them (43 per cent) has declared "not interested" or "middle". A journalistic profession in Russia is perceived as a dangerous one. Many journalists who covered political topics were murdered. Perhaps this is why the Russian respondents differ from others – it is safer to not be interested in politics and political questions.

Figure 7.3: Interest in politics and political questions in society (in per cent)

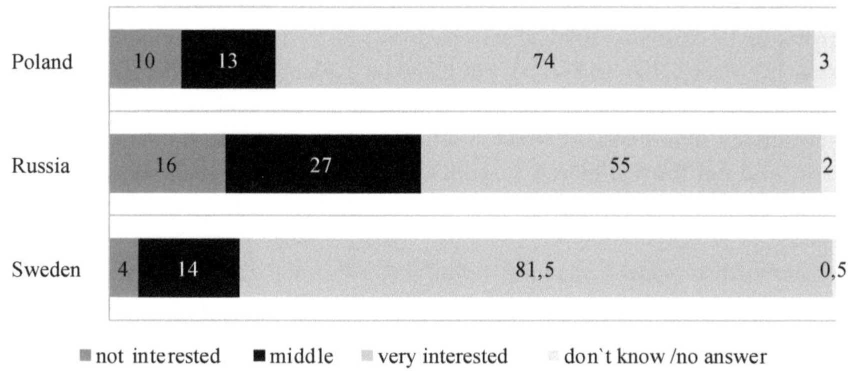

Source: *How much interested are you in politics and political questions in society?* (N 500= 100 % in the case of each country). The "Journalism in Change" Project Survey, 2012.

7.1.4 Journalistic values and politics

We asked journalists about values and duties that should be important for them. We verified some statements on journalism and ethics in the context of politics, as an acceptance of working as a journalist and to be active in politics at the same time, political influence in the media company, whether a journalist should be a neutral reporter and control public authorities (see Table 7.2).

Table 7.2: A journalist cannot himself be active in politics at the same time as working as a journalist

Country	Disagree (1 – 2)		Yes and no (3)		Agree (4 – 5)		Don't know / no answer		M (1–5)
	N	%	N	%	N	%	N	%	
Poland	38	7.6	24	4.8	432	86.4	6	1.2	4.54
Russia	162	31.4	114	22.8	208	41.4	16	3.4	3.14
Sweden	123	24.6	51	10.2	316	63.2	10	2	3.74

Source: *Here are some statements on journalism, ethics and politics. Do you agree or not? (Give your answer on a scale 1–5 disagree – fully agree).* (N 500= 100 % in a case of each country). The "Journalism in Change" Project Survey, 2012.

The biggest part of Polish respondents (86.4 per cent) has not accepted a situation when a person working as journalist could be active in politics at the same time and this is reflected by the very high mean, 4.54. This opinion was also expressed by 63.2 per cent of Swedish respondents (the mean 3.74). It can be surprising because professional standards as well as media freedom work much better in Sweden than in Poland, but it can be explained by a different historical experience of both nations – rather positive in Sweden and negative in Poland. Additionally, in Sweden the word "politics" is used and understood in a broader sense, and not limited merely to political parties. In the Polish consciousness, this same word is more connected with politicians and focused on political parties.

The Swedish media had a long tradition of political parallelism in the 19th century and also in the 20th century, when the newspapers had clear political preferences, and they were firmly anchored in political parties (Hallin and Mancini, 2004:154). According to the research of Høyer and Lorentzen (1977), party dailies represented 97 per cent in 1974. Asp (1986) claims that a symbiotic relationship between journalists and politicians were typical then. The Swedish Broadcasting Corporation passed the other way. The public television and radio stations are close to the BBC model, with "a relatively high degree of autonomy from political influence" (Hallin and Mancini, 2004:169; Humphreys, 1996:165–7). The Swedish case reflects well the exchange model of "the relationship between politicians as communicators and media personnel as observers of the political scene", which was presented by Blumler and Gurevitch (1995:29–30). Journalists have not perceived any risk from political parties and politicians in this country. There are some advantages and disadvantages of this kind of relationship. A border between two participants of a communication process is invisible and is easy to cross. If the media and journalists have a weaker position, then, party logic can dominate in the news (Mazzoleni, 1987). It seems that in Sweden politicians and journalists have this same position and they are partners in a communication process.

The Polish experience is different. During the communist times, but also after the collapse of this system in 1989, political actors tried to subordinate the media. But Polish journalists were not an easy object to control thanks to a strongly romantic idea of being free and autonomous, which was born in the 19th century, when Poland did not exist on the map. The respondents do not have any doubts in these questions, only 1.2 per cent of them do not know or they do not answer. The principle to separate the journalistic profession from political activity is expressed stronger in Poland than in Sweden. It can be explained by two factors. Firstly, there was and is a significant tradition of oppositional *samizdat* press in Poland and a historically-rooted attitude to be against ruling groups or leaders (in the 19th century when Polish territory was divided amongst Russian, German and

Austrian monarchies, and during the communist period against the Soviet regime). Secondly, it can be a result of a negative experience of many journalists in Poland during recent last decades, a lot of them lost their job or were forced to leave the profession after "an affair " with politics. In Sweden, a higher level of acceptance for journalistic activity in politics is descended probably from a strong political parallelism in the 19th century and in the first half of the 20th century, when many journalists were members of political parties or cooperated with politicians and supported political parties, but they maintained high journalistic standards.

According to many scholars, a journalistic profession in Russia is perceived as a political one (Pasti, Czernysh, and Svitich, 2012). The relationships between media and politics were and still are very intensive and deep. A deeply rooted tradition of "prophetic" and "enlightened" imperial Russia and subsequently, communist Soviet Union produced a high conformity with the reality of political instrumentalization and an evident tradition of clientelism. For more than seven decades, journalism was a profession controlled by the Soviet regime and served it (Pasti, Chernysh, and Svitich, 2012:267). Nowadays, classic political parallelism as in the West does not work. Party press still exists but it is not a powerful medium. The control of monarch and communist party was replaced by many ties of journalists with politicians and business (Bodrunova, 2013:9). For this reason, we predicted that an acceptance for an interactivity on both sides, and this kind of behavior in Russia is still high. The data collected in our project confirms this opinion. More than 31cent of respondents have not seen a problem to be journalist and be active in politics at the same time (only 7.6 per cent in Poland and 2.8 in Sweden). The significant difference among the three countries (above all Poland) is also confirmed by the study means: the highest for Poland -4.54, lower for Sweden- 3.75, and the lowest for Russia -3.14.

The next research problem concerns acceptance of political influence in the media company (see Table 7.3). Polish and Swedish respondents have quite a similar opinion. Definitely the majority of journalists in Sweden (about 88 per cent) and Poland (about 76 per cent) do not accept political influence, and only a marginal number of respondents says that it is acceptable (in Sweden -1.6 per cent, Poland – 5.6 per cent). A different standard has been observed in Russia (the highest mean among three countries – 2.59 versus 1.31 in Sweden and 1.72 in Poland). More than 44 per cent of Russian respondents perceive this influence negatively, but – and it is an interesting feature in context of our analysis – more than 20 percent accept this influence. It can be explained by the fact which t Nygren and Degtereva underline (2012), that journalists choose a job according to their political views and preferences.

Table 7.3: As a journalist one has to accept political influence in the media company

Country	Disagree (1 – 2)		Yes and no (3)		Agree (4 – 5)		Don't know / no answer		M (1–5)
	N	%	N	%	N	%	N	%	
Poland	379	75.8	71	14.2	28	5.6	22	4.4	1.73
Russia	222	44.4	153	30.6	103	20.6	22	4.4	2.59
Sweden	438	87.6	29	5.8	9	1.6	24	4.8	1.31

Source: *Here are some statements on journalism, ethics and politics. Do you agree or not? (Give your answer on a scale 1–5 disagree – fully agree).* (N 500= 100 % in the case of each country). The "Journalism in Change" Project Survey, 2012.

The answers for the question, linked with the role of the journalist as a neutral reporter, are very surprising (see Table 7.4). It is a case where researchers should interpret the data carefully. Paradoxically, in Russia, where political influence has been accepted at a higher degree than in the other two countries, the majority of respondents (85 per cent) indicate that a journalist should be a neutral reporter, and it is a little bit higher than in Poland (82 per cent) and Sweden (75 per cent). How is one to interpret this journalistic conviction in an authoritarian regime, where a freedom of media is low? Perhaps we could explain this result as a desire of the representative of this profession for the future? The other surprise arrived from Sweden – the country with the highest journalistic culture, located at the top of the Press Freedom Index. "To be a neutral reporter" has not been important for 8 per cent of respondents, more than in Poland (58 per cent) and Russia (2.6 per cent). Generally, the means in each or three states are very high (Poland – 4.26, Sweden – 4.21, and in Russia the highest one – 4.41).

Table 7.4: A journalist should be a neutral reporter

Country	Not important (1 – 2)		Yes and not (3)		Very important (4 – 5)		Don't know / no answer		M (1–5)
	N	%	N	%	N	%	N	%	
Poland	28	5.6	56	11.2	409	81.8	7	16.2	4.26
Russia	13	2.6	51	10.2	432	86.4	4	0.8	4.41
Sweden	40	8	75	15	378	75.6	7	16.2	4.21

Source: *How important are the following professional duties for a journalist in your country? (Give your answer on a scale 1–5 disagree – fully agree).* (N 500= 100 % in the case of each country). The "Journalism in Change" Project Survey, 2012.

We decided to compare the professional duty "to be a neutral reporter" with "control public authorities" (see Table 7.5). In this case we received a result which confirmed a very high journalistic culture in Sweden, where a clear separation between politics and journalism (94 per cent) is a widely accepted value. Polish journalists have expressed the same opinion (about 88 per cent). There is a visible gap between Poland and Sweden, on the one hand, and Russia, on the other (a difference of 40 per cent in comparison with Sweden and 34 per cent in comparison with Poland). Only the margined group of Swedish (1.8 per cent) and Polish (2.6 per cent), but 18 per cent of Russian respondents disagree with journalistic control over public authorities. If we add Russian respondents who choose "yes and no" (more than 23 per cent) or do not answer this question (4 per cent), a huge difference in the research means (3.66 in contrast to Sweden – 4.69 and Poland – 4.55), we can conclude that the idea of a watchdog function of the media l is still not developed in Russia due to the authoritarian regime of the Kremlin. This opinion is supported by Vartanova, who claims that the mass media "played a tacit role in the power system of Russia" (2012:134). It can be explained also by the strong position of state and its needs, where journalists traditionally were servants of political power (Trakhtenberg, 2007: 123).

Table 7.5: Journalists should control public authorities

Country	Not important (1 – 2)		Yes and no (3)		Very important (4 – 5)		Don't know / no answer		M (1–5)
	N	%	N	%	N	%	N	%	
Poland	13	2.6	33	6.6	441	88.2	13	2.6	4.55
Russia	90	18	118	23.6	272	54.4	20	4	3.66
Sweden	9	1.8	16	3.2	470	94	5	1	4.69

Source: *How important are the following professional duties for a journalist in your country? (Give your answer on a scale 1-5 disagree – fully agree).* (N 500= 100 % in the case of each country). The "Journalism in Change" Project Survey, 2012.

7.2 Obstacles at work from political actors

In this part of the analysis we are interested in the opinions of respondents about their connections with political actors – politicians and political parties, governments and state authorities. We asked them about:

- factors influencing the selection of subjects in daily work of the media organization (political connections and strong political actors outside the media company);
- obstacles before the final publishing or broadcasting (political interest of the media pressure from political actors outside the media company);
- evaluation of the influence of the state (laws and ownership) and political actors in the media company, and its change during the last 5–10 years;
- appreciation of press freedom level in the country.

In general, journalists who cover politics should have a better knowledge about this area and more contacts with politicians than other representatives of this profession. It can be very important to know that there were only a few respondents between 2.6 to 4.4 per cent in the three countries, who covered politics as a special topic (13 Poles, 22 Russians, and 17 Swedes). In fact, our data came from journalists who are not experts in political issues, but all of them are citizens and work for media companies.

7.2.1 Factors influencing the selection of subjects in daily work

We were interested in the relationship of politicians and journalists. For this reason we asked the respondents about the significance of political connections (affiliation) of the media organizations and the role of strong political actors outside of the media in daily work. The answers for these two questions can help us to appreciate not only how the democratic standards are adopted by the news media, but also, say more about the democratic culture of political actors.

We predicted that in Sweden the affiliation of these two participants of political communication process – political actors and media organizations – has not been a significant factor. The data confirm this hypothesis (see Table 7.6). About 77 per cent of Swedish respondents declare that it is an "insignificant" factor, and the high political parallelism belongs to the past. The low mean (1.58) and low standard deviation (0.951) indicate that the data points tend to be very close. According to many indexes and rankings, Swedish democracy is one of the most solid in the word.

In the case of Poland, we observe a balance among those journalists who selected "insignificant" (35 per cent), "yes and not insignificant" (22.6 per cent) and "significant" (25 per cent), and is confirmed by the mean (2.78). But the standard deviation (1.310) is high, and the highest among the three states, which indicates that the data are spread out over a large range of values (see Table 7.6). Polish journalists had a negative experience with political affiliation of media companies, above all with public broadcasting. This disapproving evaluation of political affiliation was strengthened by a quite fresh experience with the presidency of Aleksander Kwaśniewski (1955–2005), the government of Jarosław Kaczyński (2006–2007) and the presidency of Lech Kaczyński (2005–2010). Recently, we observe that media organizations try to avoid political connections and tend to be independent.

In Russia, the biggest part of respondents (42 per cent) claims that the selection of subjects in the daily work of their media organization is effectuated under "significant" influence of political connections. The mean is relatively high (3.26) and higher in comparison with a low mean for Sweden (1.58) and also Poland (2.78).We note that the standard deviation is lower than in Poland, which testifies that Russian journalists are more homogenous in their opinion. Other indexes and rankings show that the relationship of political actors and media organizations in Russia are typical rather for the exchange model with a weak media position (Blumler and Gurevitch, 1995:27–31). It means that the media organizations are linked strictly with political power, and they are subordinate to the Kremlin. Even if independent media exist in the state, the majority of respondents work for state media. One possible explanation is that Russian respondents are not interested in politics as a specific topic (only 4 per cent cover politics and less than 1 per cent – foreign news), and they do not feel any political pressure, and they do not have direct contact with political actors. It is the area reserved for their managers, editors and directors.

Table 7.6: Significance of political connections (affiliation) of the media companies in the work in news organizations

Country	Insignificant (1 – 2)		Yes and no (3)		Significant (4 – 5)		Don't know / no answer		SD	M (1–5)
	N	%	N	%	N	%	N	%		
Poland	177	35.4	113	22.6	125	25	85	17	1.310	2.78
Russia	127	24.5	132	26.5	208	42	33	7	1.251	3.26
Sweden	384	76.5	45	9	29	5.5	45	9	0.951	1.58

Source: *Different factors influence the selection of subjects in the daily work of your media organization. How would you evaluate the significance of the following factors on work in your news organization? (Give your answer on a scale 1–5 insignificant– very significant)* (N 500= 100 % in the case of each country). The "Journalism in Change" Project Survey, 2012.

The political temperament of Poles and a clear lack of acceptance of these practices can be explained by historical context. Swedes and Russians never lost their states and independence. Their governments, monarchs, party secretaries were their own, not from outside. They created a strong attitude against an official hostile power and they did not accept an affiliation with a "repressive" political system. The oppositional movement was strong, as in the case of "Solidarity". A lot of journalists worked for *samizdat* press or oppositional media, which were published outside the country. Many journalists, who worked for the official media, developed perfectly a skill to write "between the lines" and the Polish public could indeed read between the lines. Polish journalists, as underlined by Curry (1990), were strongly active politically but at the same time, they developed a high journalistic culture. Polish journalists played a fundamental role in the process of development of national identity. In this specific situation, only politically involved journalists were able to maintain hope of independence in society.

Swedish journalists, despite a strong political parallelism in the 20th century and the first half the 20th century, created solid professional standards. They have not perceived political affiliation as a significant obstacle, because they have not had a negative experience with a political instrumentalization of the media in the past as Poles. Russia is a more complicated case to explain. In the country with a communist and a consequent strong authoritarian regime experience, characterized by deep politicization and treatment of the media as a political tool, which should support the ruling elites, journalists have not evolved a sufficient sensitivity for political connection of the media. They either cannot evaluate the significance of this factor, or they do not like to do it.

There is no difference between Polish and Russian journalists in the answers to the question about the influence of strong political actors outside the media company (see Table 7.7 and Chapter 5.4.1). About 4 per cent fewer Russians than Polish respondents think that it is an "insignificant influence", and about 10 per cent more Russians appreciate it as "significant". The means are quite similar – for Poland 2.44 and a little bit higher in the case of Russia – 2.79. Swedes do not note this problem in their country – more than 68 per cent of respondents claim that this kind of influence is "insignificant" and the mean (1.82) is substantially lower. This data confirms the general knowledge about professionalism and democratic standards characterized a relationship between journalists and politicians in the three states. In Sweden, through a long and strong period of political parallelism in the past, each professional group preserves its autonomy and independence typical for two equal partners. Polish journalists, have experienced the communist period, then a time of transition toward democracy and democratic consolidation, had a strong tradition of political involvement in the democratization process. Nowadays, they do not agree with the significant and influential role of political actors, who still have ambitions to control the work of news organizations, above all public radio and television. Journalists working for public and private broadcasting have a different experience and it is reflected in a high standard deviation. Additionally, about a quarter of respondents do not know how to answer this question. The changes of relationship between politicians and journalists after the Kaczyński period are still very dynamic and maybe they could find a suitable reply.

The interpretation of Russian data in the case of Russia seems to be the most difficult. The mean is a little bit higher, but a standard deviation is less than in Poland. More than 26 per cent of respondents evaluate the role of strong political actors as being "significant". Perhaps journalists who work for state-owned media, do not note this influence or they appreciate it as something evident, which is typical for the adversarial model (Blumler and Gurevitch, 1995: 27–31).

Table 7.7: Significance of strong political actors outside the media company in the work in a news organization

Country	Insignificant (1 – 2)		Yes and no (3)		Significant (4 – 5)		Don't know / no answer		SD	M (1–5)
	N	%	N	%	N	%	N	%		
Poland	203	40.6	102	20.4	77	15.4	118	23.6	1.270	2.44
Russia	184	36.8	132	26.4	131	26.2	53	10.6	1.234	2.79
Sweden	341	68.2	70	14	34	6.8	55	11	1.003	1.82

Source: *Different factors influence the selection of subjects in the daily work of your media organization. How would you evaluate the significance of the following factors in the work in your news organization? (Give your answer on a scale 1–5 insignificant– very significant.* (N 500= 100 % in the case of each country). The "Journalism in Change" Project Survey, 2012.

7.2.2 Obstacles before final publishing/broadcast

Obstacles which appear before final publication and broadcast of journalistic materials are one of the essentials for the relationship between media and politics (see Figure 7.4). Taking into consideration a high journalistic and political culture in Sweden, we predicted that this problem did not exist in Sweden or it had a marginal relevance. The data confirmed this presupposition. More than 87 per cent of respondents have never or rarely met in their professional career an obstacle such as political interests of the owners of the media company. In the case of Poland, more than 53 respondents have chosen this answer. The problem is more visible in the context of Russia, where 24 per cent of respondents claim that it happens often and very often. The difference among the three states is confirmed by the means (a very low for Sweden -1.21, and higher for Poland -2.23 and Russia -2.69) and standard deviation (0.523 for Sweden, 1.194 for Poland and 2.69 for Russia).

Figure 7.4: Political interest of the owners of the media company as an obstacle before the final publishing broadcast (in per cent)

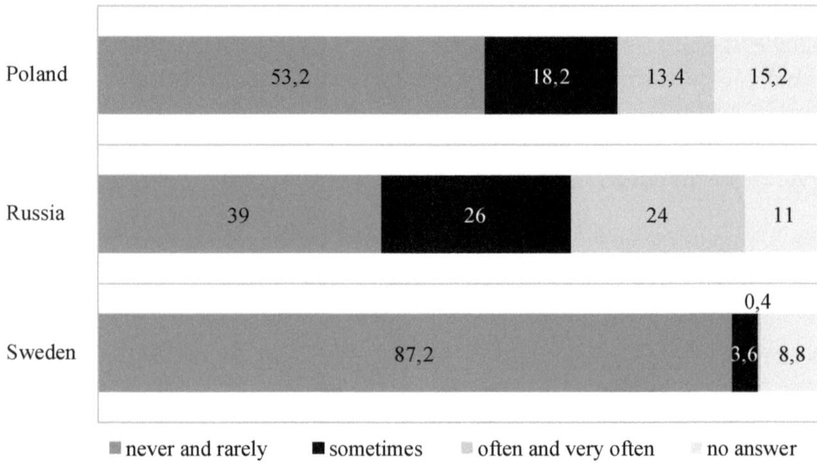

Source: *A journalist can face various obstacles before the final publishing/broadcast. How often do journalists at your media company face pressure from political actors outside the media company? (Range your answers on a scale 1–5).* (N 500= 100 % in the case of each country). The "Journalism in Change" Project Survey, 2012.

The question about pressure from political actors outside the media company as an obstacle before the final publishing or broadcasting journalistic materials is one of crucial importance in the context of this research (see Figure 7.5). As in the case of the previous one, we predicted that in Sweden it would not be a problematic issue. Eighty-five per cent of respondents never or rarely have met this obstacle in their career. In Poland, this obstacle does not exist in the opinion of 59 per cent of all respondents, but in this country the largest group has not answered this question (about 19 per cent). In Russia, about 43 per cent of respondents claim that they have never faced any pressure from political actors outside the media company. But in this country, the group of respondents who noted these practices, is the largest (18 per cent vs. 1.6 per cent in Sweden and 8.6 per cent in Poland). In this context, we observe a significant difference between Sweden, on the one hand, and Poland and Russia on the other. In the case of Sweden, this data confirm once again that the overlapping of the spheres of politics and journalism does not exist or is very limited, and journalists and politicians function separately.

Figure 7.5: Pressure from political actors outside the media company as an obstacle before final publishing/broadcasting (on per cent)

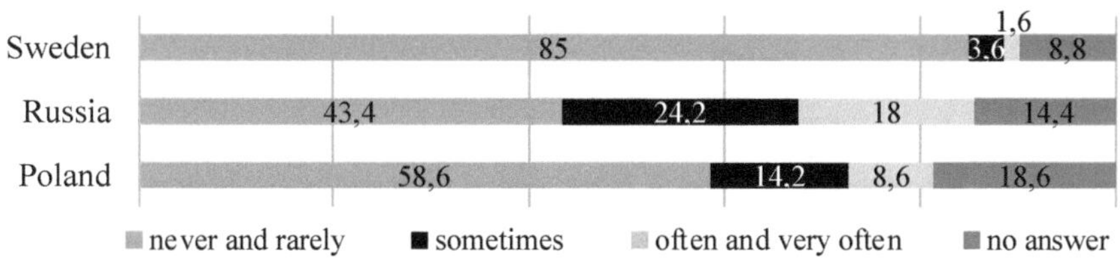

Source: *A journalist can face various obstacles before final publishing/broadcast. How often do journalists at your media company face pressure from political actors outside the media company? (Range your answers on a scale 1–5).* (N 500= 100 % in the case of each country). The "Journalism in Change" Project Survey, 2012.

Analysis of the data from the different media sectors (see Table 7.8) shows that political pressure is not evaluated in the same way by the respondents. In Poland and Russia, journalists employed by the public (in Poland) and state (in Russia) radio and television, additionally in magazines, have faced this obstacle more frequently than in other sectors. Summing up, we can conclude that this factor is absolutely unimportant in Sweden, it happens in Poland, and it is quite problematic in Russia.

Table 7.8: Pressure from political actors outside the media company in media sectors

Media sectors	Poland						Sweden						Russia					
	N each sector =100%		Never and very rarely (1–2)		Often and very often (4–5)		N each sector =100%		Never and very rarely (1–2)		Often and very often (4–5)		N each sector =100%		Never and very rarely		Often and very often	
	n	%	n	%	n	%	n	%	n	%	n	%	n	%	n	%	n	%
National dailies	51	10.2	24	4.8	2	0.4	67	13.4	55	11	0	0	110	22	41	8.2	19	3.8
Regional dailies	81	16.2	45	5	7	1.4	143	28.6	121	24.2	1	0.2	63	12.6	26	5.2	18	3.6
Magazines	127	25.4	76	15.2	10	2	94	18.8	81	16.2	1	0.2	153	30.6	77	15.4	21	4.2
Public radio	64	12.8	36	7.2	8	1.6	42	8.4	38	7.6	2	0.4	7	1.4	5	1	0	0
Private radio	40	8	27	5.4	2	0.4	3	0.6	3	0.6	0	0	18	3.6	7	1.4	0	0
Public television	46	9.2	23	4.6	9	1.8	37	7.4	35	7	1	0.2	43	8.6	15	3	10	2
Private television	52	10.4	37	7.4	2	0.4	20	4	18	3.6	0	0	38	7.6	13	2.6	9	1.8
Media online	29	5.8	17	3.4	3	0.6	8	1.6	8	1.6	0	0	40	8	21	4.2	6	1.2
Press agency	10	2	8	1.6	0	0	21	4.2	19	3.8	0	0	10	2	4	0.8	3	0.6
Other	0	0	0	0	0	0	65	13	47	9.4	3	0.6	18	3.6	8	1.6	4	0.8
Total numbers	500	100	293	58.6	43	8.6	500	100	425	43.4	8	1.6	500	100	217	43.4	90	18

Source: *A journalist can face various obstacles before final publishing/broadcast. How often do journalists at your media company face the following obstacles? (Range your answers on scale 1–5)* (N 500= 100% in each country). The "Journalism in Change" Project Survey, 2012.

7.2.3 Evaluation of the influence of state and political actors

The independence of journalism can be influenced by many factors. Some of them come from the political system and are generated by the state and political actors. The state institutions, such as parliament, create the laws that constitute the main frame for media systems. Additionally, governments can apply norms and regulations in each country in different ways. Political institutions decide about an audiovisual council, print press regulations or ownership of the media. As a consequence, they can limit the independence of journalism in formal (law and regulations) and informal (pressure, friendship) ways. Polish (about 52 per cent) and Russian (about 60 per cent) journalists are well aware of this danger, due to their negative experiences in this area. Only 25 per cent of Swedish respondents indicate this factor. The difference among the three countries is expressed by the means – a higher one in Poland (3.45) and Russia (3.78), and one point less in Sweden (2.74). Additionally, we note that a standard deviation in each state is high, which t illustrates distinct differences among the respondents (see Table 7.9).

Table 7.9: The independence of journalism can be influenced by state influence (laws and ownership)

Country (1–5)	No danger (1 – 2)		Yes and no (3)		A great danger (4 – 5)		No answer	SD	M (1–5)
	N	%	N	%	N	%	N		
Poland (n 448=100 %)	81	18.1	135	30.1	232	51.8	52	1.092	3.45
Russia (n 491=100%	67	13.6	132	26.9	292	59.5	9	1.119	3.78
Sweden (n 460=100%)	198	43	147	32	115	25	40	1150	2.75

Source: *The independence of journalism can be influenced by different factors. How do you evaluate the influence for the following factors? (Give your answer on a scale 1–5: no danger– a great danger).* The "Journalism in Change" Project Survey, 2012.

Political influence is still perceived by journalists as one of the most perilous for press freedom and professional autonomy. But we did not offer any definition of this kind of influence and each respondent could understand it differently according to her/his personal experience and knowledge. We predicted that this problem could be absent or very narrow in the case in Sweden. The data support this opinion. Polish respondents have stressed this problem in the highest degree (about 68 per cent). We can explain it by a quite fresh experience with the

Kaczyńskis' rule and a large need for journalists to be independent. Only 12 per cent fewer Russian journalists express this same belief (about 56 per cent). The difference among the three countries is reflected by the means: no danger between Poland (3.88) and Russia (3.65), much lower in Sweden (2.43). In this case we note also a quite high standard deviation in each country (see Table 7.10).

Table 7.10: The independence of journalism can be influenced by political influence in media companies

Country (1–5)	No danger (1 – 2)		Yes and no (3)		A great danger (4 – 5)		No answer	SD	M (1–5)
	N	%	N	%	N	%	N		
Poland (n 455=100 %)	52	11.4	93	20.4	310	68.2	45	1.033	3.88
Russia (n 485=100%	84	17.3	131	27	270	55.7	15	1.221	3.65
Sweden (n 461=100%)	247	68.2	141	55.7	73	15.8	39	1.073	2.43

Source: *The independence of journalism can be influenced by different factors. How do you evaluate the influence for the following factors? (Give your answer on a scale 1–5 no danger– a great danger).* The "Journalism in Change" Project Survey, 2012.

The problem of the independence of journalism is reflected by the media content. We asked the journalists how they evaluated the changes in influence on this scope during the last 5–10 years (see Figure 7.6). In the three countries comparable numbers of respondents are convinced about "decrease" and that it is "the same". A significant difference has been noted in the category 'increased'. Only 3.4 Swedish respondents have chosen this answer but more than 30 per cent of Russians and about 23 per cent of Poles. It means that in post-communist countries this problem is still visible and journalists are afraid of this kind of limitation of their professional activity. Additionally, a lot of respondents in each state have not answered this question – still 38 per cent of Swedes, 27 per cent of Poles and about 22 per cent of Russians. It suggests that they do not have any experience in this aspect, perhaps it is a problem of managers and editors and it does not touch the average journalist. If we compare this data with the answers to the question dedicated to the independence of journalism vs. the influence by political influence in media companies (see Figure 7.5), we note that the means are lower (the highest one for Russia – 3.27, a little bit lower for Poland at 3.14, and lower for Sweden – 2.62), but Swedish and Polish journalists do not differ in their opinion (the standard deviation for Poland – 0.980, and for Sweden – 0.779), but in the case of Russia it is significantly high (1.919).

Figure 7.6: The changes in influence on media content by politicians (in per cent)

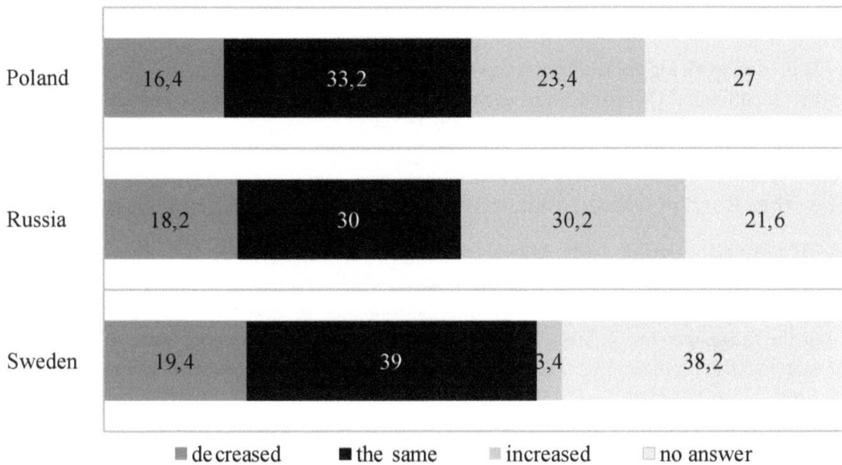

Source: *If you look back 5–10 years, how would you evaluate the changes in influence on media content in your new organization from the following factors? (give your answers on a scale 1–5).* (N 500= 100 % in the case of each country). The "Journalism in Change" Project Survey, 2012.

However, if we compare the general opinions with the responses in particular sectors in which the respondents are employed, then the picture is a bit different. In the case of Poland, an increase in the influence of politicians has been noted by more than 18 per cent of journalists employed in national dailies and more than 30 per cent in magazines. In the case of these two sectors the data once again confirm the close connections between the world of politics and the media, and the high political involvement of editorial offices. The same conclusions concern public service television – more than 18 per cent of choices are for "increase" and only 2 per cent for "decrease" of influence. Russian journalists employed by the national (30 per cent), regional dailies (36 per cent), magazines (30 percent), state and private television (39 per cent) have claimed that influence on media content from politicians increased.

During in-depth interviews we asked journalists about their experience with political influence and pressure, about owners and their political orientation. There were seven Russians and only four Swedes, who wanted to share their opinions about relations with the political world. However, they answered briefly "No, I don't feel any pressure", "Not at all", "No, that's not present", "We do not feel any influence from owners and their political preferences". All of them – Russians and Swedes, have not noted any political pressure from outside. In the case of Russia, the opinions presented by journalists in the interviews contradict the survey results. One Russian said that:

> "I used to work for politically committed media. Nowadays this magazine has no political "inclination." Of course, there are employees who do not hide their political preferences, but it in no way affects work." [RR 17, male, 49 years old, editor-in-chief of a web-based media, special correspondent of magazine]

The other Russian talked an internal political pluralism in the medium:

> "The magazine permits itself political criticism. Sometimes we present conflicts between different political positions in our pages. There is conflict of political preference between employees, which does not prevent any of us from expressing our opinions in the pages of the magazine. The publishing house has no influence on our preference of subject matter." [RR 15, male, 35–50 years old, head of the department, national weekly magazine]

Only one journalist noted that he was not permitted to criticize the ruling elites in Russia:

> "Our editor's office has a certain political tone and, accordingly, a certain political climate. Even though I work in the neutral social area, I cannot promote liberal ideas due to my personal beliefs and general course of the newspaper (…). *Naturally, none of us attacks the ruling party and the powers that be* [BD-O]." [RR12, female, 28 years old, head of department, national weekly magazine]

The Swedish interviewers were sparing in their opinion about pressure from political actors. It has not been an important question at all for them. We can explain it by a lack of this problem in Sweden thanks to a long democratic experience, a strong professional culture, and a sense of journalistic independence, but also by a high culture of political actors, who respect traditionally a journalistic autonomy. Paradoxically, they have complained about insufficient contact with politicians:

> "Nowadays, they rarely have direct contact with politicians, it is often the press secretary that calls here. In the past, politicians might call directly (…) they called almost daily, now they go through a press secretary. It was easier before, straight shooting. When you interview actors today it's like a circus, it is well – arranged today. They choose who gets to be in. It is stricter." [SR 1, male, 56 years old, reporter, daily newspaper]

> "Many [politicians –BD-O] call a reporter that they have contact with. Some know what we might be interested in. Others get in touch and are angry or happy afterward. Then it's often up to me. "Should you not do this instead, that was crazy," this also applies to politicians, they will occasionally send emails. Then you try to settle things." [SR 2, female, 37 years old, radio reporter]

Polish journalists basically differ from Swedish and Russian interviewees. They have talked a lot and they have given many examples. Additionally, what has not happened in the case of respondents from the other two countries is that Poles have expressed some extreme opinions. On the one hand, journalists have not seen any problem in their contact with politicians. They have not noted the obstacles or political pressure from political actors or owners who have been involved politically:

> "(…) if any political party does something wrong, makes a mistake, we always write about it. I have never had the impression that somebody has botched something, but because we like them, we don't write about that (…) [PR2, female, 30 years old, reporter, national newspaper]

In the case of Polish media owned by international media groups, they have known nothing about the political preferences of the owners.

> "No, no. The owners are German and I haven't noticed any influence. To tell you the truth, I don't even know what their political orientation is." [PR7, female, 35–50 years old, editor, a local daily].

These kinds of comments come above all from journalists who worked for the print media or private radio and television stations. The journalists employed by public media companies have perceived the situation in a different way. Some of them have criticized strongly the practices in public radio and television:

> "Especially when it is public radio or television, the current political option becomes so important, so visible." [PR13, female, 35–50 years old, freelancer]

> "It has happened (…) there were phone calls after the publications and not some interventions beforehand. Nobody has ever forced me or forbidden me to write something." [PR16, female, under35 years old, editor of the portal of public radio]

One journalist, who has worked for public television for 20 years and was thrown out of it twice, has observed various mechanisms from a middle-level managerial position, which was a consequence of managers' formal nomination[2]. He claimed:

> "I am convinced that the system doesn't change, whichever political party is in power at a particular moment. What's more, any person removed from the management of public television, usually seeks revenge at their earliest convenience. This really happens. And every time it gets worse. It's a phenomenon that winds itself up, from what

2 Members of the Audiovisual Council are nominated by political institutions (President, Sejm and Senate). The managers of public radio and television are nominated by the Audiovisual Council.

> I've observed. It's like a political hamster, running around and producing revenge after revenge. This of course eliminates any kind of independent journalism, because you become independent when they sack you (…)." [PR5, Female, 50 years old, public TV, manager]

These citations support qualitatively the data that we have received in the survey – one country and two other visions of the same profession and practices, which are linked with the personal experience of journalists who have been employed in the public and private sectors. It is an argument which can also support Jakubowicz's thesis about a correspondence between the political system and PSB in post-communist countries in the 2000s described as "systemic parallelism" (2008:41).

7.2.4 Perception of the level of press freedom

Finally, we asked journalists about changes in their perception of press freedom in the last 5 to 10 years (see Figure 7.7). Thanks to the *Press Freedom Index 2012*, we know (see Chapter 2.1.3.) about a huge gap in this ranking between Sweden (12) and Poland (24), on the one hand, and Russia (142), on the other. The survey conducted in this same year showed that about 47 per cent of Russian respondents are convinced that press freedom "decreased" and is twice as high in the other two countries. It means that this serious problem is noted by representatives of the journalistic profession. The difference between Poles (about 26 per cent) and Swedes (about 24 per cent) is not significant. The media system in Sweden is the most stable, not only among the three states but also in the world. This opinion is expressed by the respondents – about 44 per cent of them conclude that press freedom stays the same. In Russia (about 32 per cent) and Poland (about 34 per cent) journalists think like that. We discover an interesting characteristic in the case of Poland. The data published by the *Press Freedom Index* from 2002 to 2014 showed huge changes (see Figure 2.2). Poland was in 58th place in 2006 in the rankings and it was the worst result during the last decade. Since then, this position has improved systematically to 19th position in 2014. This process was noted by more than 31 per cent of Polish respondents – three times more than in Sweden (about 10 per cent) and about twice that of Russia (about 17 per cent).

Figure 7.7: Press freedom in the opinion of respondents (in per cent)

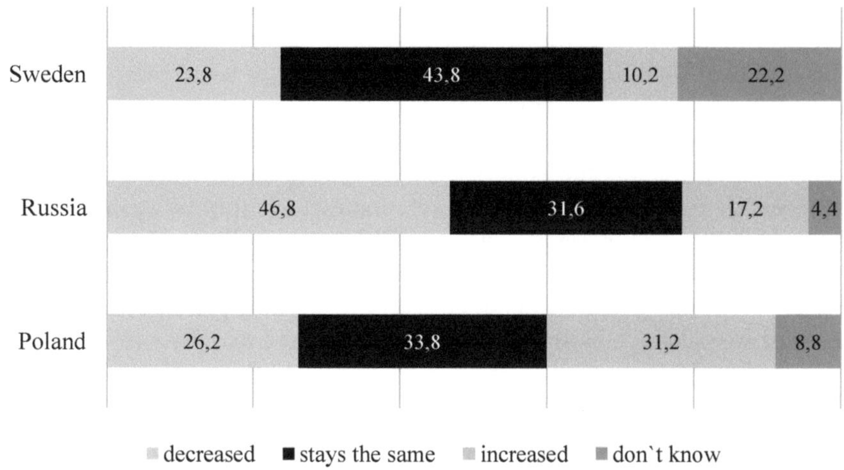

Source: *Do you think the level of press freedom in your country has increased or decreased during the last 5–10 years* (N 500= 100 % in the case of each country). The "Journalism in Change" Project Survey, 2012.

7.3 Summing up: The relationship between political actors and journalists

At the beginning of this chapter the hypothesis has been presented that political systems and democratic standards determine the journalistic culture in each country and produce different models of relationship between media and politics. We have tried to verify how democratic values and everyday practices of politicians and journalists have worked in light of the empirical data, which we received by conducting quantitative surveys and qualitative interviews.

It is not easy to compare three countries with such different historical, religious, cultural and economic backgrounds. One common feature is the location around the Baltic Sea. Two of them have played a hegemonic role in the regions, Sweden in Scandinavia and Russia (Soviet Union) in Central and Eastern Europe, when a mass press rose in the 19th century. Poland as a state did not then, exist and its territory was shared by three European superpowers. Swedes created a solid democracy, while Russians cultivated the idea of an imperial superstate, effectively eliminating freedom inspirations of conquered nations such as Poles. Nowadays, we can find some features linked with these historical conditions. Swedish and Russian journalists have acted in their national state and they identified with it. This was a situation totally different in the case of Poles, who struggled for a long time with the foreign and hostile political power of three great monarchies in the 20th century and Soviet regime in the 20th century. This historical condition created one specific attitude of Polish journalists – to be against the official authorities. This feature we discover also in the project's results. Polish journalists who participated in the research are the most critical towards political actors and express strongly their aspiration of being independent. Swedish journalism elaborated its autonomy many centuries ago and nowadays journalists do not perceive it as a problem. The collapse of communism and breakup of the USSR opened some new possibilities for Russian journalists. However, they present a rather critical opinion about political standards in the state nowadays. Almost half of respondents noted that press freedom had decreased in the country and a third of them claimed that influence on media content by politicians had increased.

Democratic values are strongly underlined by Polish respondents, because they still feel insecure and are distrustful even 25 years after the collapse of the communist regime. Probably for this reason, they claim that the influence of political connections of media owners and political actors outside the media company is one the most significant factors which threatens ajournalistic autonomy. The difference between Poland and the other two countries is significant in this aspect. Polish journalists are able to separate much stronger than Swedes and Russians journalistic work and political activity. On the one hand, they do not accept politics as an alternative profession, and are strictly against working as a journalist and being active in politics at the same time. On the other, we observe a strong identification of journalists with political ideologies and values (but not with political actors).

Swedish journalists do not emphasize political interest of media owners and pressure from political actors outside the media company, which can be explained by a long democratic tradition and high political culture and civil liberties (*Democracy Index*, 2012). Polish journalists, on the one hand negatively experienced this, and having a high democratic ambition, are afraid of political influence, but on the other hand, they have observed political interest of media owners and pressure from political actors only on a limited scale. Experienced by the practices of the Kwaśniewski and Kaczyński presidencies, they perceive to a higher degree than others a great danger of political influence in media companies.

Finally, we try to compare the journalists' responses with the *Press Freedom Index* and *Democracy Index*. The next surprise is that Polish journalists, frustrated by and dissatisfied with the relationship between media and politics, showed the most positive picture of press freedom among the three counties, and were indicated as the smallest group who said that press freedom had "decreased" and the largest that it had "increased".

The survey confirms some fundamental differences in connection with the level of democracy in the three countries (see Figure 7.8).

Figure 7.8: Democracy Index 2006–2012 (overall score)

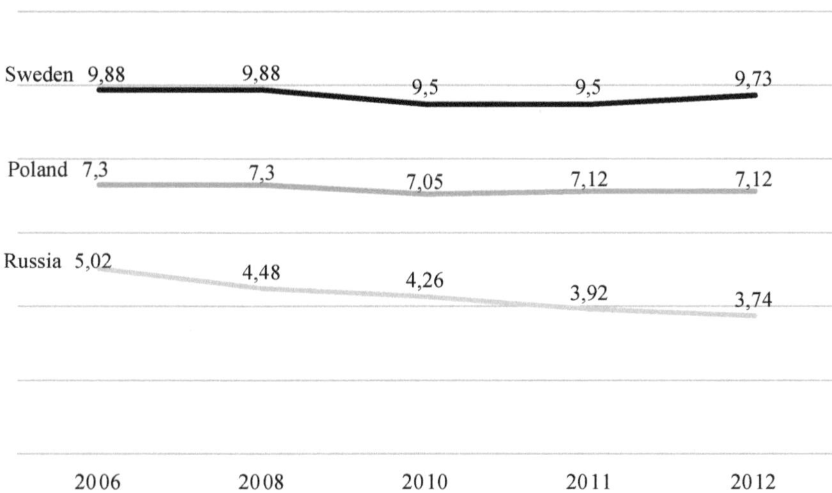

Source: http://www.economist.com/node/2006,2008,2010,2011,2012.

Sweden was located in the *Democracy Index* in second position among the "full democracies" in 2011 and 2012.[3] Poland was classified as a "flawed democracy" and improved its rank from 46th in 2006, 45th in 2011, to 44th in 2012.[4] In addition, the authors of this index note that the regional average score for Eastern Europe declined in 2012 compared with 2011. This process is generated above all by

3 The overall score in the case of Sweden was better in 2012 than in 2011 (+0.25 points) thanks to higher notes in the category scores as political participation (8.89 points in 2011 and 9.44 points in 2012) and political culture (8.89 points in 2011 and maximum 10.00 points in 2012). The results in electoral process and pluralism (9.58 points), function of government (9.64) and civil society (10) were the same in 2011 and 2012.

4 In Poland, a value of each category source was the same in 2011 and 2012: electoral process and pluralism –9.58, functioning of government –6.44, political participation –6.11, political culture –4.38, civil liberties –9.12.

Russia being classified as an "authoritarian regime" in this Index because democratic standards are not respected there on a visible scale. In one year this state fell from 117th in 2011 to 122nd place in the rankings in 2012[5] (*Democracy Index, 2012*).

Journalistic culture in Sweden is strong and typical for a full democracy. Journalism in Poland is in a time of a critical moment. It changes deeply and quickly, but our data support that democratic standards in Poland do not work in the same way as in Sweden, and we can still count this state as a flawed democracy. Russia differs in many aspects from Sweden and also Poland. The answers of respondents let us claim that the board between politics and journalism in Russia is much weaker than in the other two countries. More than half of the surveyed Russian journalists accept politics as an alternative profession. These data can be supported by the opinion of Pasti, Chernysh, and Svitich who consider journalism as "a social lift" (2012:275). Russian respondents accept political influence at a higher level than others and significantly, fewer of them do not see a need for the control of public authority. These data are not sufficient to claim that Russia is an authoritarian regime, but they let us formulate an opinion that journalistic culture in this aspect is significantly weaker than in Sweden and in Poland.

Our data confirm that the different models of media and politics relationship can be applied in each of the three countries (see Figure 7.9).

Figure 7.9: Models of relationship between media and politics in Poland, Russia and Sweden

← partnership model -- domination model →

SWEDEN	POLAND	RUSSIA
the symmetrical model	the flawed symmetrical model	the asymmetrical model

Source: Author

In the case of Sweden, it is the classical partnership model of exchange (Blumler and Gurevitch, 1995:66–67) with symmetrical relations between politicians and journalists (Dobek-Ostrowska, 2004:221). Despite a long tradition of the party press, this model was developed and consolidated due to dialogical democracy,

5 Russia lost – 0.20 points in the overall scores. The most complicated situation is observed in political culture (from 3.13 points in 2011 to 2.50 in 2012) and civil liberties (from 4.71 points in 2011 to 4.41 in 2012). Other category sources, such as electoral process and pluralism, (3.92), functioning of government (2.86) and political participation (5) were also appreciated as low.

high civic culture and participation of citizens, strong journalistic autonomy and political actors respected the independence of the media. Additionally, about 62 per cent of Swedish respondents are more than 40 years old. It lets us claim that they are experienced and they express traditional values and attitudes created by many journalistic generations.

The standards typical for Polish journalism were created in totally different historical conditions, which resulted in the asymmetrical model with the dominance of politicians. But this model was challenged strongly by difficult competitive relations with journalists, who were politically involved and, in the majority of cases, against the official authorities. Successive journalistic generations had to confront with successive political pressure of the strong power of the Prussian, Russian and Austrian empires, and consequently, the Soviet regime. However, Polish journalists were not an easy object of political manipulation. They had strong ambition to be independent (despite their visible political values) and they stood up to political actors who wanted to instrumentalize the media. About 20 percent of respondents, who are more than 45 years old, started to work in this profession before 1989, and work until today. In this case we can talk about a relative equilibrium among generations. Nowadays, the flawed symmetrical model of relationship media and politics is typical for Poland, and is characterized by the strong confrontation of journalists and politicians, and the passive role of weak civic society structures.

Russia, as an authoritarian political regime, represents the asymmetrical model of media and politics relationship with a strong domination of respective Tsars and Communist party's secretaries in the past and President of the state today. On the one hand, Russian journalists, who work for the official media nowadays, support the political regime, and this opinion of a young female web reporter confirms it:

> "We need to remember that the policy of media defines all in the daily work of the journalist. Every journalist chooses his/her place of work." (Nygren and Degtereva, 2012:6)

On the other hand, in contrast to Sweden and Poland, this profession was (and still is) an extremely dangerous one for opposition journalists and bloggers, who act against the official power (Pasti, Chernysh and Svitich, 2012:278). The reports of *Russian Union of Journalists* (2011) and *Deaths of Journalists in Russia* (2014), (see also *56 Journalists Killed in Russia,* 2014) present the data of the violent, premature or unexplained deaths of more than 300 journalists in Russia since the early 1990s.

Nowadays, the majority of critical journalists work above all for online media or newspapers and magazines with a limited circulation. This kind of media is popular rather in the biggest cities and they are consumed by a relatively low number of recipients. Research conducted by HSE showed that inhabitants from rural parts of the country have access to new technology and digital media but they use them rather in a passive way (Novikova and Kiriya, 2013; Kiriya and Novikova, 2013; Kiriya, 2013). They prefer to watch state television channels and use digital media for better quality. This phenomenon is described by Kiriya and Novikova as "depressive media consumption" (2013), and is explained as a mental problem, which leads to the information ghetto, and produces a habit to absorb propaganda passively (Kiriya and Degtereva, 2010).

The mainstream media, which are present everywhere in this huge country, are owned and controlled by the government authority or big business (Kiriya and Degtereva, 2010). Perhaps the Russian sample is not representative for all journalists in this country. About 70 per cent of the respondents are under 36 years old and they cannot remember the communist period, or they were very young. Only 7 per cent of respondents are older than 60 years old. It tells us about a lack of continuity and a significant exchange of journalistic generations. Young journalists have not had a possibility to develop a journalistic ethics (Pasti, Chernysh and Svitich, 2012:274), a professional autonomy and journalistic culture. The data show that Russian respondents in general accept this situation. They are not interested in politics a lot, they agree with many practices of the owners and do not complain about a lack of autonomy or political pressure (Nygren and Degtereva, 2012). It may be that they are not able to discover these kinds of practices or they believe that a conformist attitude as safer.

We should remember that media systems are very fragile and dynamic structures. The data which are presented above reflect the quality of relationship between politics and journalists in the beginning of the 2010s. We cannot predict whether the same conclusions will be valid well in few years. Paradoxically, it seems that the most important risk for the quality of journalism would arrive not from the political sphere, but economic systems. In the case of Russia, it is not only a danger from the authoritarian regime of the Kremlin, but also the interests of political and economic powers, which act together and are strictly linked. We will search these connections in the next chapter.

8. Journalism and commercialization

Bogusława Dobek-Ostrowska

Analysis of the economic context of the mass media systems and finding a place for journalism and journalists in this environment is the main aim of this chapter. Today there is no doubt that media companies are part of the business world and they should accept the rules operative in this world. For this reason, the answers to these questions could help in this research process:

RQ1: Does economic pressure exist in the media market today?
RQ2: What kind of external economic pressures on the media market are present in each country?
RQ3: Do advertisers limit or support autonomy and quality of journalism?
RQ4: Is internal economic pressure noted in Poland, Russia and Sweden and, if yes, how does it work?
RQ5: How does economic interest of media companies determine the quality of journalism?
RQ6: Are foreign owners present in the country and what role is played by them?
RQ7: How do journalists perceive the changes in influence on media content from the economic market?
RQ8: What do they think about the changes in influence on media content from the owners during the last decade and what do they predict in the future in this aspect?

The research questions provoke the following hypothesis:

H: *Commercialization is in progress in each of the three countries, only the speed and the consequences of this process are different.*

8.1 The media in pincers of profits

Until the 2000s, traditional communication research has not dedicated a lot of attention to the economy and economic context of the mass media (Doyle, 2002b:1), which were perceived as a fundamental element of the public and political sphere in the 19th and 20th centuries in Europe. Even if they belonged to private owners (as the print press), public and political interests were the most important and a discussion about economic profits was rarely undertaken. The process of the media industry's formation was reinforced by a launch of private broadcasting media (or privatization of state TV and radio stations) in the 1970s and 1980s. In the 20th century, it was subsequently marked by the rapid development of the Internet and easy access to new communication technologies in the 1990s. Croteau and Hoynes claim that the 21st century began with a bang and noted:

> "New Technologies are helping to transform the 21st century media industry into something quite different from its predecessors. Although these changes suggest exciting possibilities, they also raise troubling questions about ownership and control of information, about concentration of power, about the responsiveness of media to an increasingly diverse society, about the future of journalism, and about the relationship between media and citizens." (Croteau and Hoynes, 2006:2)

They analyzed two models – the market and public sphere models. They have no doubt that the first one was more expansive. It takes successively more place and limits the second. This situation has produced two conflicting logics, which are typical for the mass media today. In the market model, "commercial" media logic dominates because private media companies are perceived as producers of goods and the main aim is to sell their products. In this context, the primary purpose of the media is to generate profits for owners and stockholders. The people who read, listen to, watch and use the media are treated as consumers, not as citizens and recipients of the news. The mass media encourage them to "enjoy themselves, view ads and buy products" (Croteau and Hoynes, 2006:39). The public sphere model, with non-commercial media, is in real danger.

This new situation in the media market pushed the traditional press and broadcast media towards commercialization. Anderson and Ward (2007:39) define this phenomenon as a new technique of sale, where profits and market value of the medium are more important than journalistic ethics or social function of media.

Commercialization is defined as a process, which leads the mass media to a state in which the structure and contents reflect the profit-seeking goals of firms and their owners, and when the media are governed by market conditions. It is today a crucial problem for the quality of the mass media. McQuail (2001:107, 492) lists a lot of negative consequences of this process as: trivialization expressed in low cultural quality, exploitation of 'weaker' consumers, utilitarian and calculative relations, propaganda for consumerism, commodification of culture and relations with the audience, and also sensationalism and personalization, commodification, standardization. This process promotes less creative content and a lack of the independence from their owners and advertisers.

In Western Europe, a process of commercialization was opened basically by strong competition. Commercialization strengthens systematically in many countries, even where there the tradition of public sphere values, a high journalistic professionalization, media accountability and civic responsibility were deeply rooted, as in Democratic Corporatist model countries, and Sweden is a good example. The difference is only in the speed of this process. However, it has brought not only negative connotations. Thanks to it, recipients have access to more products of different quality and they can select among them. Swedish media and its consumers still maintain high standards, and the state is noted in this context as one of the best in the world (see *Democracy Index*, *Press Freedom Index*).

Central and Eastern Europe were touched by commercialization after the collapse of communism at the beginning of 1990s and it developed quickly. Privatization was a first step to commercialization. The negative consequences of this process have been revealed in a short period of time, above all the low quality of media content. But on the other hand, thanks to privatization, the offers of the media are richer and consumers have more choice. But looking for profit by many owners has changed the media market visibly during the last two – three decades (see Figure 8.1).

Figure 8.1: Phases of media commercialization in post-communist Europe and its consequences

PRIVATIZATION

↓

COMPETITION

↓

OWNERSHIP CONCENTRATION

(economic instrumentalization)

↓

CONTENT EFFECTS

(tabloidization, infotainment, politicotainment, newszak)

Source: Author

Privatization, as mentioned above in Poland and Russia, was the first step towards the consequences of commercialization. The press sector was the first one to open in all the countries in Central and Eastern Europe, and also the first one to suffer the effects of this process. Competitiveness of the new markets and lack of national investors, who could invest in this business, attracted foreign owners. The big and rich media companies, above all from Germany and Switzerland, began to invest in the print press market. Soft regulation and lack of proper anti-monopoly laws made this region attractive for the expansion of foreign capital. Powerful Western media groups, such as Swiss Ringer or German Axel Springer,

Bauer, WAZ, Verlagsgruppe Passau, appeared immediately. They bought existing titles and offered a lot of new ones. Commercial broadcasters entered a little bit later, but they needed only a little time to play a dominant role in the market (Dobek-Ostrowska, 2008:17–18). Poland, as the biggest market in the region, differs from the others. Polish national companies invested in radio and television stations. Privatization not only heralded a long-expected process of media pluralism and also competition. By contrast to other states in Central and Eastern Europe, a balance in the market among public and private broadcasters has been noted. Nowadays, strong competition is observed in each media sector in Poland. Traditionally, the public broadcasting service has had a privileged position in Sweden. The print media market is competitive, but public subsidies for an editor of the second daily national newspaper on the market still exist, and it is this which significantly differentiates this Scandinavian state from Poland and Russia. Competition is visible but gentler than other European countries. In the case of Russia, the situation seems to be different. According to Pasti, Chernysh, and Svitich, Russian media belong directly or indirectly to the government or government-controlled organizations:

> "Nearly 80 % of the press consists of non-market publications that are affiliated closely with financial – industrial groups and partially serve as a cloak for business, or state-owned organizations with financing from regional and local budget." (Pasti, Chernysh and Svitich, 2012:267)

Political connotations and linkage between political elites and oligarchs are the most important. The broadcasting media belong to the state and state-business companies such as Gazprom Media, which co-owns the state-commercial mixed TV channels. In this way, the state indirectly controls the most popular media. The mass media are instrumentalized by both political and economic powers, which have different aims but they achieve them in a way of "cooperation". On the one hand, the ruling group helps businessmen be richer, and, on the other hand, oligarchs support political elites to keep the power (Vartanova, 2012: 133–134).

Concentration of ownership is a typical consequence of privatization. Central Europe is a very good example of how pluralist markets gradually change into an oligopoly or even a monopoly. For example, the regional daily newspaper sector is dominated by one owner, Verlagsgruppe Passau in Poland and Czech Republic. Ownership concentration leads to the transnational ownership of regional networks and cross- media ownership. Cross – media ownership is especially visible in Poland, which is the biggest market in Central Europe, where vertical and horizontal concentration meet (Dobek-Ostrowska, 2008:19). An economic calculation results in that one media group has more newspapers and magazines, more television and radio stations, because it reduces costs and make profits, which is the most important aim of business. In the case of Poland, a trend towards greater media concentration takes a form of market colonization by internal and external (mainly from Germany) media groups. This trend is much weaker or does not exist in Sweden and Russia. But in each of the three countries, the owners desire to generate higher profits. The media companies limit their expenses, which causes a reduction of employment and lower salary to journalists. As a consequence, consumers receive a lower quality of media content, which is produced for sale.

The pressurized economic world has profoundly changed traditional journalism during the last 10 to 20 years. Commercialization is responsible for a gradual elimination of social values and standards, and displacement of the public sphere model (Croteau and Hoynes, 2006:39). This phenomenon is reflected by the media content. Contemporary journalism gives in to tabloidization and bases itself on trivialization of content, above all in the Liberal model of media system and new democracies like Poland. It means that the media concentrates on easy and attractive news. This kind of media content is called infotainment or politicotainment. In this context, a conscious neologism *newszak* (news + muzak[1]) introduced by journalist Malcolm Muggeridge and developed by Franklin (1997: 4) is very useful. This concept shows how news values and news formats used by the media changed in the 1980s and 1990s. It reflects the new trends in journalism, as a replacement of hard news by soft news, "stories which interest the public above stories which are in the public interest", as sensations and scandals with celebrities in the roles of heroes, crime stories in place of international affairs (Franklin and others, 2012). A focus on entertainment rather than serious information is fateful. It has brought infotainment and politicotainment, which displace systematically professional journalism.

The media and journalism are in the pincers of money. Many scholars claim that profits of owners and advertisers are more important than serving the public interest, promote active citizenship or social integration (Jakubowicz, 2011:161–2). Some of them indicate "de-professionalization" and "proletarization" of journalism as a visible consequence of commercialization in this part of Europe (Volek, 2010:188). However, advertisers play an ambivalent role in the media market and their relationship with media companies can take a different form. Advertisers and media companies frequently have opposing interests. Not every

1 'Muzak' is recorded music that is played quietly and continuously in public places, such as airports, hotels, and shops, to make people feel relaxed.

time do advertisers exert pressure on the media, sometimes the media companies do it and try to. The first wants to pay less for advertising, the second looks for profits and charges high prices. Thanks to advertisers the media can exist and develop their activity, enrich their content and offer. Both sides need each other.. But, the need of profit from advertisers can constitute a danger for the quality of media. The producers of goods and services pay to international or local advertising agencies, which create, plan and handle advertising for clients. These agencies can offer additional services and forms of promotion for their clients from business, corporations, governments or non-profit organizations. Sometimes, it is not enough for the sponsors of advertisements and then they try to exert pressure directly on the media.

In this chapter we would like to present the opinion of our respondents about commercialization. How do they perceive this process? When do they see the economic risks and how do they evaluate them? What kind of threats – external or internal – are more dangerous in the opinion of journalists? What role is played by economic power, advertisers and private business outside media companies, and how do they influence media content? How do the respondents evaluate the interests of media companies, and their demand for profit and foreign ownership?

8.2 External economic pressure on the media market

In this part of our research we concentrate on the media market and external environment of each media organization regardless of its size. There are many sources of economic pressure as other media companies act in the same or different media sector, which generates competitive reactions. Economic pressure arrives from various industries and business institutions. The crucial place is occupied by advertisers and sponsors with their money that flows in the media market.

8.2.1 Economic power outside the media companies

We asked respondents to evaluate a significance of economic power outside the media company in the selection of subjects in the daily work (see Figure 8.2). The data confirmed our supposition that in Sweden this problem is perceived as quite insignificant. It is 61 per cent of respondents (the mean 2.0). This answer was chosen by 45 per cent of Russians (the mean 2.52) and 31 per cent of Polish journalists (the mean 2.80) (compare Chapter 5.5.1). In the case of Poland, we note not only the highest mean among the three countries but also the standard deviation – 1.308 vs. 1.270 in Russia and 1.118 in Sweden, which confirms a significant dispersion from the average. Polish journalists feel the significance of economic power outside the media more than others (only 4 per cent more in comparison with Russians but 15 per cent with Swedes. Additionally, the biggest group of them (about a quarter) do not know how to reply, which suggests that they do not have sufficient knowledge about relationships with business companies and advertisers because it is the task of media managers, owners or editors. They are unsure in this changeable situation and they do not have an opinion.

Figure 8.2: Evaluation of the significance of economic power outside the media company (in per cent)

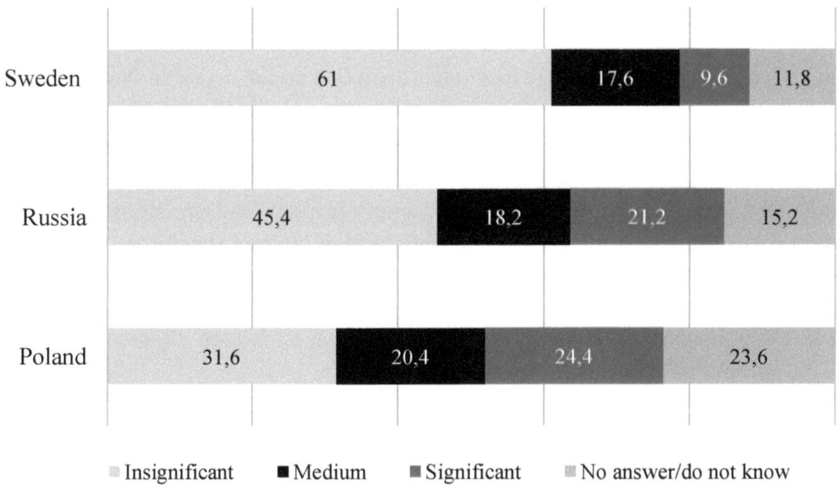

Source: *Different factors influence the selection of subjects in the daily work of your media organization. How could you evaluate the significance of the following factors in the work on your news organization? (Give your answer on a scale 1–5)* (N =500 in each country). The "Journalism in Change" Project Survey, 2012.

In this context, a question linked with the changes in influence of private business on media content during the last 5 to 10 years seems justified (see Table 8.1). Generally, it was one of the most difficult questions for the respondents. A lot of journalists do not reply to it: more than 40 percent of Swedish and about 32 per cent of Polish and Russian respondents. We can try to explain it by a lack of experience in the case of Russia because the majority of respondents (69 per cent) are younger than 35 years old (compare Chapter 3.2, see Figure 3.3) or they have not had any contact with private business. But it is rather difficult to use this argument in the case of Swedish and Polish journalists. More than 29 per cent of Polish respondents are convinced that this influence has increased. A comparable group of Swedish (21.8 per cent) and Russian (11 per cent) journalists select this answer but it is lower than in Poland. The biggest number of Russian respondents believe that influence had decreased (32 per cent) vs. limited groups of Polish (about 7 per cent) and Swedish journalists (11 per cent). These data in the case of Russia can suggest that journalists do not know a lot about this kind of relationship and they do not perceive a risk from private business.

This difference among the three countries is reflected by the means: the highest in Poland – 3.41, in Sweden – 2.93 and the lowest in Russia – 2.55. This quite significant difference among the countries is surprising. The next time, the data show that Polish journalists do not feel comfortable with the economic market and private business, which limits media independence. Additionally, a significant part of the media belongs to foreign owners who are active in Poland above all to make a profit. In the case of Sweden, we can explain it by a high journalistic culture on the one hand, and a strong public service and high standards of business organizations, which refrain from influencing media content, on the other. In Russia, probably the journalists have a very limited relationship with private business or they do not have it at all. This kind of contact is reserved for media owners and editors and is invisible for media workers and public opinion.

Table 8.1: Evaluation of the changes in influence of private business on media content in the last 5–10 years

	Decrease (1–2)		Medium (3)		Increase (4–5)		No answer/do not know		SD	M
	N	%	N	%	N	%	N	%		
Poland	37	7.4	158	31.6	147	29.4	158	31.6	0.901	3.41
Russia	160	32	117	23.4	64	12.8	159	31.8	1.130	2.55
Sweden	55	11	189	37.8	55	11	201	40.2	0.853	2.93

Source: *If you look back 5–10 years, how would you evaluate the changes in influence on media content in your news organization from the following actors? (Range your answers on a scale 1–5).* (N =500) The "Journalism in Change" Project Survey, 2012.

8.2.2 Influence of advertisers

The next time, Polish journalists present the most critical opinion, when they evaluate the significance of advertisement sponsors from business and industry in a process of selection of subjects in daily work (see Figure 8.3). Fewer than

24 per cent of them indicated a role of "*those who advertise in the media*" as insignificant and more than 38 per cent as significant. About a quarter of Polish respondents do not answer this question or they do not know how to reply. Swedish journalists do not have a problem with the judgment. The influence of institutions, which pay for advertising, is noted as insignificant by more than 67 per cent of respondents. Russian journalists perceive this situation a little bit better than the Polish one but more than 31 per cent of them think that this influence is significant. This difference among the three groups is reflected by the means: the highest one for Poland – 3.30, 2.95 for Russia, and the lowest one for Sweden – 1.87.

Figure 8.3: Evaluation of the significance of those who advertise in the media in selection of subjects in daily work (in per cent)

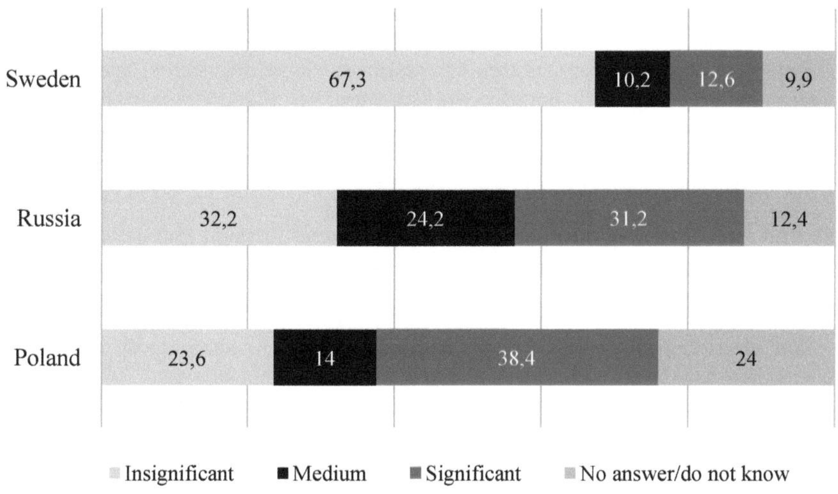

Source: *Different factors influence the selection of subjects in the daily work of your media organization. How could you evaluate the significance of the following factors in the work on your news organization? (Give your answer on a scale 1–5)* (N =500) The "Journalism in Change" Project Survey, 2012.

Analysis of the twenty interviews with Swedish journalists confirms the general opinion that this problem is still unimportant in Sweden but journalists are conscious that this process has appeared [SR15].

> "There is a wall of steel. The few times it has happened that someone has come and wondered something, they will be immediately sent to the editor. It is rare but there have been disputes recently as the advertising department is pissed because we have

> written something very lousy about an advertiser." [SR15, male, 49 years old, magazine, reporter]

The majority of interviewed Swedish journalists downplay this problem but they see that advertisers are around them on the media market and take systematically more place [SR 10].

> "The paper needs all the advertisements it can get to keep the issues and profits up. And advertisers are demanding more and more to advertise." [SR10, female, 55 years old, magazine, reporter]

In Poland, the media manage themselves with this kind of pressure in different ways. If they have a good position on the market, they are able to defy businessmen and big corporations [PR 3].

> "Are the journalists under pressure sometimes? (…) I'm not able to esteem it. I remember only one or two such actions, such scandals. Once we had a scandal concerning Kulczyk[2], then we were pressed not to put the screws on him. We resisted. But Warta and some firms of Kulczyk resigned their advertisements. So we described it. And he was also scared and came back. But there was such a text "Me, Kulczyk" with his photo on the cover. This is what I remember – but luckily 'under' this editor-in-chief and with this conventionality of our weekly, I can say that we can afford courage." [PR3, male, 55 years old, weekly magazine, journalist]

We find two opposing attitudes among interviewed Russian journalists. Few of them do not have contact with an advertising office and they do not know a lot about influence from *those who advertise in the media* [RR11]:

> "The advertising office is on a different floor, I don't even know where exactly. We work separately, have no relations with them. I came to them just once for a logo (…)." [RR 11, male, 24 years old, network of television channel, producer and correspondent]

The other journalist shows it as something obvious [RR7]:

> "The advertising office does have quite influence. It is completely natural, because advertising is one of the most important sources of income for a regional TV channel. There are cases, when we broadcast large amounts of advertising material even in prejudice of the information body of the program." [RR 7, male, between 35 to 55 years old, local TV company, group editor, anchor]

We can predict what the Swedish journalists think about pressure from advertisers (see Figure 8.4). More than 81 per cent of respondents have not faced any obstacles from them in journalistic professional work. This situation is more

---

2 Jan Kulczyk is one of the richest Polish businessmen, the owner of Kulczyk Holdings and the international group Kulczyk Investments.

complicated in Poland and Russia. In each of the two countries, 44 to 45 per cent of respondents indicate "*never or rarely*", in Poland 22 per cent and in Russia 19 per cent "*often and very often*". The difference between Sweden and the other two states is reflected by the means: the lowest one for Sweden – 1.38, higher for Russia – 2.48, and the highest one for Poland – 2.58.

Figure 8.4: Frequency of pressure from advertisers before final publishing/broadcast (in per cent)

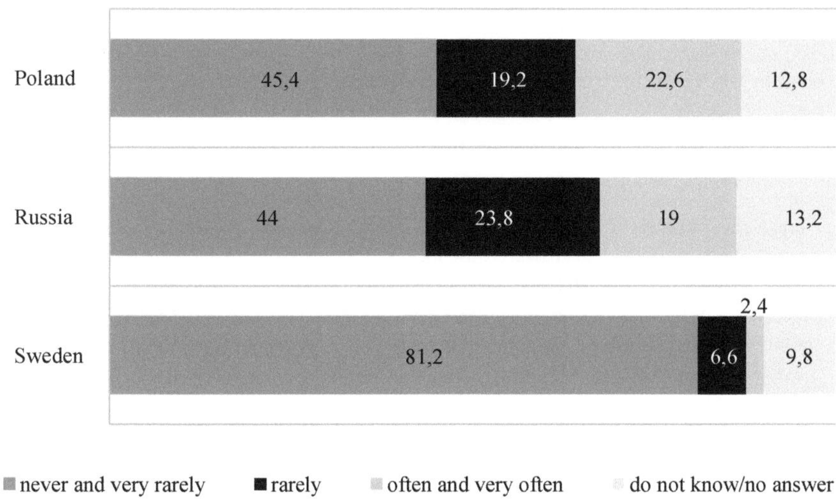

Source: *A journalist can face various obstacles before final publishing/broadcast. How often do journalists in your media company face the following obstacles? (Range your answers on a scale 1–5)* (N =500). The "Journalism in Change" Project Survey, 2012.

Polish journalists feel more painfully the pressure from advertisers than other respondents. This problem touches above all small media companies, but not exclusively. More than 64 per cent of Polish respondents claim that advertisers' influence on media content is a danger for the quality of journalism, and is 20 percent more than in Russia and twice that of Sweden. The means for Poland (3.87) and Sweden (2.91) show well this significant difference between these countries (see Table 8.2).

Table 8.2: Evaluation of advertisers' influence on content

	No danger (1–2)		Yes and no (3)		Danger (4–5)		No answer/do not know		SD	M
	N	%	N	%	N	%	N	%		
Poland	40	8	105	21	322	64.4	33	6.6	0.930	3.87
Russia	132	26.4	137	27.4	218	43.6	13	2.6	1.155	3.30
Sweden	186	37.2	114	22.8	164	32.8	36	7.2	1.057	2.91

Source: *The independence of journalism can be influenced by different factors. How do you evaluate the influence of the following factors? (Give your opinion on a scale 1–5).* (N =500) The "Journalism in Change" Project Survey, 2012.

This trouble with advertisers was described by one of the interviewed Polish journalists [PR 18]:

> "It gets more and more dependent every day. I don't really want to say but the situation in my newsroom is quite difficult at the moment because the advertising department keeps losing touch with the very thin advertising market in Poland and its focus on magazines published in great numbers; my magazine isn't really published in great numbers, therefore we have fewer advertisers, but a lot depends on them, because I have to write something. I suspect that the article that I'm writing about the Polish National Bank originates in some kind of a sponsor agreement, but I'm not sure (….) Sometimes, I'm 100% sure that it's a so-called 'sponsored article'. If I know that it's sponsored, of course I write it but you have to put a note to say it's a sponsored article. Sometimes the sponsorship is hidden or constitutes a beginning of a future relationship with a particular institution." [PR18, male, over 55 years old, journalist, weekly magazine]

The respondents were asked about the changes in influence on media content from advertisers during the last 5 to 10 years (see Table 8.3). As we can guess, Polish journalists are the most worried. More than half of them believe that this influence increased. This same opinion is expressed by 24 per cent of Russian and only 15 percent of Swedish journalists. This deep difference is reflected by the means – 3.82 for Poland and lower in the case of Sweden – 3.01 and Russia – 2.97. The data confirm that the Polish media system is more commercialized than the Russian and Swedish ones.

Table 8.3: Evaluation of the changes in influence on media content from advertisers during the last 5–10 years (in per cent)

	Decrease (1–2)		Medium (3)		Increase (4–5)		No answer/ do not know		SD	M
	N	%	N	%	N	%	N	%		
Poland	31	6.2	92	18.4	252	50.4	125	25	0.930	3.83
Russia	127	25.4	131	26.2	122	24.4	120	24	1.155	3.01
Sweden	69	13.8	157	31.4	79	15.8	195	39	1.057	2.97

Source: *If you look back 5–10 years, how would you evaluate the changes in influence on media content in your news organization from the following factors? (Range your answers on a scale 1–5)* (N =500) The "Journalism in Change" Project Survey, 2012.

There is one interesting opinion about the potential role of advertisers expressed by a Swedish interviewed journalist [SR1]:

> "Fifteen years ago it was unthinkable to talk to someone in the advertising department. But there is no impact to write in a certain direction, except in advertising attachments such as "travel in Sweden." Then you actively go out looking for advertisers, but it is handled well. The attitude has changed towards the others, people are not seen as plague-stricken, to put it bluntly. It's not as much of an anti- advertisement feeling anymore. You talk to each other. We have come closer to each other. It's still a business. But it is an approach without direct consequences for journalism." [SR1, male, 57 years old, newspaper, reporter]

In Poland, we find different visions of this problem. A journalist claims that it does not exist at all [PR 7]:

> "Advertisers? I have no idea. I don't know because I think that it is a question of negotiations, deliberations, sometimes soothing such a person – I simply don't know because it is totally beyond me. I have never been interested in the matter." [PR7, female, 50 years old, a local daily newspaper, editor, journalist]

Another respondent [PR 11] explains that the influence of advertisers is sometimes a necessity, because the medium needs money for its activity:

> "They have a lot of influence, it is all interrelated, that's why it is great that finally it starts to take shape. In the past it was complete guerrilla warfare – a journalist had to do everything himself/herself, but now fortunately, there are the departments that you are talking about, i.e. we have the advertising department, the marketing department, it is obvious what they do because of their names, and advertising is even separated, on the radio it is separated with the sounds, and in the press it is in different rubrics, right? In this way you know that it is an advertisement and you kind of know how to approach it. You know that it is not a case, it is not information. So here in radio, advertising also

means earning money for the Radio's activity, and not only for our own promotion". [PR11, female, under 35 years old, local public radio, journalist]

8.3 Internal economic pressure

External economic pressure can generate or reinforce an existing internal one (Croteau and Hoynes, 2000:58–63). The process of commercialization has passed quickly at the beginning of the political and economic transformation after 1989 in Poland. Nowadays, it is a competitive market and the media should fight about their survival every day. It seems that economic instrumentalization is more advanced there than in the two other countries which participated in the project, but the reasons for this situation are dissimilar in Sweden and Russia. In Sweden, where the media system is characterized by a strong professional culture and non-profit public service sector, commercialization is restrained by media policy and the welfare state, which supports these journalistic standards. We can explain it also by one of the highest levels of readership, not only in Europe but also in the world that holds back a crisis in the print press. However, circulation of newspapers is also declining in this country. The media ownership structure in Russia is quite different from Poland and Sweden, and it produces different economic conditions. The main traditional media belong to the state, commercial companies, or state and commercial capital. The private media and the press published by foreign owners play still a marginal role. The media market is not as competitive as in Poland. In the Russian case, the media are often supported by the capital of their owners.

8.3.1 The owners' demand for profits

In general, private commercial and non-commercial media cannot exist without a profit. From this point of view their incomes are crucial for the development and activity on the market. Predominantly, the negative tendency is observed when the profit interests of owners override audience interest and quality of content, and the competitive economic market starts ruling. The owners' demand for profit is perceived as dangerous for journalistic independence by all participants of the survey. According the previous answers, Polish respondents expressed the most critical opinion (see Table 9.4). About 77 per cent of them think that it is a danger. It is 10 per cent more in comparison with Sweden and 16 percent more in comparison with Russia. The mean in case of Poland is also the highest, 4.13 vs. 3.86 in Russia and 3.75 in Sweden.

Table 8.4: Evaluation of the influence of The owners' demand for profit on journalistic independence

	No danger (1–2)		Medium (3)		A danger (4–5)		No answer/ do not know		SD	M
	N	%	N	%	N	%	N	%		
Poland	27	5.4	73	14.6	386	77.2	14	2.8	0.920	4.13
Russia	40	8	124	24.8	331	66.2	5	1	1.015	3.86
Sweden	69	13.8	90	18	305	61	36	7.2	1.183	3.75

Source: *The independence of journalism can be influenced by different factors. How do you evaluate the influence of the following factors? (Give your opinion on a scale 1–5)* The "Journalism in Change" Project Survey, 2012.

In this context, it seems very important what one interviewed Swedishjournalist said:

> "(…) nowadays the purpose is that owners would make money, once upon a time it was about creating good journalism with the money." [SR6, male, 46 years old, daily newspaper, reporter]

In general, Swedish journalists have not had contact with the owners and they have not suffered by their demands for profit.

> "The owners, Bonnier. I have not had much to do with them at all. It's a long way between me and those who sit and decide." [SR 13, female, 42 years old, daily newspaper, reporter]

We find two kinds of comments among interviewed Polish journalists. In general, all of them have seen the danger of the owners' demand for profit, but they had other experiences with it. The journalists who were employed by small media [PR 9], they felt this pressure stronger than those who worked for a bigger one [PR8, PR 10].

> "Of course, it is influenced (states firmly). It is a system of "communicating vessels". It is influenced as we work in close cooperation with the advertising department, it is especially true about me in my capacity as editor and supervisor. So I need to know, for example, which events are organized under our auspices (…) It affects us for sure because it is obvious that we are a local station and we need to earn our keep. So I need to know what kind of events they are, how they are going to be advertised, if we advertise them one week before the actual event because that's what we usually do, as there are kind of established rules and we stick to them (…)." [PR9, female, under 35 years old, commercial local radio station, reporter]

> "(…) pressure from the owners? Yes (…) It has happened. I think it happens… not very often but it happens. It shouldn't have but it happened. Sometimes some reports had to be re-worded or they couldn't be broadcast at all because of so-called pressure from the top. But these are rare cases."[PR8, female, 35–55 years old, public television, reporter]

> "I don't feel the pressure that we talk or hear about from time to time, I don't feel the pressure that somebody comes up to me and says: "you mustn't write about it, about that case, about that person because they are our advertisers". No, never. During my career, already a long one, it has happened maybe three times, in newspapers where I worked that such situations happened and they were quickly straightened out because my media organization cares about its reliability and it won't be reliable and trustworthy if it doesn't raise some issues only because of advertising." [PR 10, female, 55 years old, regional daily newspaper, journalist]

Russian journalists have not dedicated a lot of attention to owners and advertisers in the interviews. The majority of them talked in the interviews about public relations and press relations. The perception of owners' influence by Russian journalists has not been the same, and it depended on the size of media where they have been employed. We find two different opinions. One was expressed by an experienced journalist from a niche national newspaper, who has seen this influence as something obvious [RR19], and the other by a young journalist, who was employed by a citizens' television channel [RR20].

> "Of course, there is some influence—first of all, by the advertising office. It all depends on us. There are situations when the issue has already been laid out, then the advertising comes and demands cutting the article. At the same time, we are all well aware that advertising is what feeds us. We have to turn the heads on and to compromise, to negotiate. Do some rotation of texts, sometimes postpone something or remove the material, so as not to spoil the text." [RR 19, female, 50 years old, niche national newspaper, editor-in-chief]

> "The heads of those offices, advertising services sometimes tried to press, but due to lack of professionalism they only thought about the fact of advertising, not the way material was presented (…) We do not feel special influence from the finance office. The business subdivision lives its own life, I haven't been there in a long time." [RR20, female, 25 years old, citizen's private television channel, information, broadcast producer and journalist]

Some journalists perceive advertising incomes as something natural and they accept all forms of cooperation with advertisers and other institutions, which offer additional profits [RR13].

> "Timesof product promotion involves good earnings. I do not refuse the use of press releases, as well as proposals to write an article from advertising executives, for it brings good income. And it may not necessarily be an offer from a newspaper; I also accept offers from other organizations, occasionally helping in the preparation of promotional leaflets and flyers." [RR13, female, 57 years old, federal newspaper, journalist]

One of the most significant consequences of commercialization is the shrinking size of editorial staff. This trend is most visible in Sweden and Poland where a clear majority of journalists have this experience (see Chapter 4). Other visible effects of economic instrumentalization and pressure of "profits" are changing working conditions, the growing role of advertisers and sponsors who send additional promotion materials or finance of special inserts, blurbs, etc. The respondents and interviewed journalists mentioned the "system of communicating vessels" or "blurring of the lines between newsroom and advertising offices."

8.3.2 Influence of foreign ownership

After the collapse of communism in Poland, the media market was able to develop due to foreign owners, above all in the sectors of private radio, regional daily newspapers and magazine sectors. Thanks to them it was possible to create the pluralistic and modern media market. They introduced a lot of new good quality products. Companies from outside Poland until today play an important role and they take a much bigger part of the market share than in Russia or Sweden. For this reason, Polish respondents have more experience with foreign owners – both positive and negative. However, they feel more at risk than other journalists: 36 per cent vs. 26 per cent in Russia and 22 per cent in Sweden (see Table 8.5). The means confirm (the highest in Poland – 3.2, and lower 2.8 in Russia and Sweden) that in Poland the process of commercialization has brought economic mechanisms which were unknown earlier. Two decades after the privatization of the media market in this state, foreign ownership is strongly rooted, and is reflected in the opinion of respondents.

Table 8.5: Evaluation of foreign ownership influence on journalistic independence

	No danger (1–2)		Yes and no (3)		A danger (4–5)		No answer/ do not know		SD	M
	N	%	N	%	N	%	N	%		
Poland	124	24.8	155	31	181	36.2	40	8	1.233	3.20
Russia	190	38	145	29	133	26.6	32	6.4	1.242	2.81
Sweden	169	33.8	129	25.8	110	22	92	18.4	1.128	2.75

Source: *The independence of journalism can be influenced by different factors. How do you evaluate the influence of the following factors? (Give your opinion on a scale 1–5).* The "Journalism in Change" Project Survey, 2012

Probably for this reason, only the Polish interviewed journalists mentioned foreign owners, no one from the Russians and Swedes did this. One of the Poles talks about the Bauer Group, which is one of the biggest companies on the Polish market, the owner of the first radio station and the most powerful editor of magazines:

> "We do get pressurized, but by the chairman not by the owner – Bauer – who's in Germany and I don't even know if he listens to our radio, maybe he doesn't; we have absolutely

no contact with him, every now and then we get some corporate bulletins and that's it. Now, the chairman! The managers meet up with him and the journalists simply don't get involved". [PR 15, male, up to 35 years old, national private radio station, journalist specialist in business news]

Polish journalists do not meet owners directly. They have contact with the managers (chairmen) who act in the name of the owners and must ensure a profitable business. It is not easy to multiply profits and create independent and professional journalism at the same time.

8.3.3 Economic interest of media companies

Economic interest of media companies can be achieved in many ways and can produce some negative consequences as obstacles before the publishing or broadcast. The answers which we received in the survey, show that there exists a difference between Sweden, on the one hand, and Russia and Poland, on the other (see Table 8.6). More than 71 per cent of Swedish respondents never or very rarely have been faced with economic interests of the media company, which is more than twice in comparison with the answers of Polish (34 per cent) and Russian (29,8 per cent) journalists. This significant difference between Sweden and the other two countries is supported by the means – 1.71 for Sweden vs. 2.91 in case of Poland and 3.01 – Russia. The answers for this question, show us a very high professional culture of owners and managers in Sweden, who separate definitely the managerial tasks and economic interest of the company from journalistic activity. In general, it is not a problem for the majority of Polish and Russian respondents (1–3 about 57 per cent).

Figure 8.6: Frequency of being faced with economic interest of the media company as obstacles before final publishing/broadcast

	Never and very rarely (1–2)		Rarely (3)		Often and very often (4–5)		No answer/do not know		SD	M
	N	%	N	%	N	%	N	%		
Poland	170	34	119	23.8	142	28.4	59	11.8	1.208	2.91
Russia	149	29.8	137	27.4	168	33.6	46	9.2	1.233	3.01
Sweden	358	71.6	59	11.8	38	7.6	45	9	1.060	1.71

Source: *A journalist can face various obstacles before final publishing/broadcast. How often do journalists at your media company face the following obstacles? (Range your answers on a scale 1–5).* (N = 500). The "Journalism in Change" Project Survey, 2012.

8.3.4 The changes in influence on media content from owners during the last 5–10 years

The question dedicated to the changes on media content from owners for the last 5 to 10 years seems to be one of the most difficult in the entire survey for respondents – 38 per cent of Swedish and a quarter of the journalists from Poland and Russia do not answer and a comparable number of respondents (about 30 per cent) from the three countries indicate that it is "medium" (see Table 8.7). The difference among the means for the three states is not as high, 3.18 for Sweden, and 3.67 for Poland and 3.53 for Russia, but when we analyze the data linked with "decrease" and "increase" of those changes in influence on media content, we note that the three groups of journalists do not perceive it equally. Only 4.4 per cent of Polish journalists evaluated those changes positively (influence on media content from owners decreased), but 40.4 per cent negatively (36 per cent difference), and it is a huge difference in comparison with Sweden, where it is inconsiderable (about 5 per cent difference – 12.4 vs. 17.2 per cent) and Russia (11 per cent difference – 16.2 vs. 27.8 per cent). These data show that the changes in Polish journalism are more dynamic and deeper, and the role of media owners is more visible and dominant. Unfortunately, this question is general and does not let us evaluate the character and quality of those changes in influence.

Figure 8.7: Evaluation of the changes in influence on media content from owners during the last 5–10 years

	Decrease (1–2)		Medium (3)		Increase (4–5)		No answer/do not know		SD	M
	N	%	N	%	N	%	N	%		
Poland	22	4.4	149	29.8	202	40.4	127	25.4	0.924	3.68
Russia	60	16.2	127	30.4	204	27.8	109	25.6	0.106	3.53
Sweden	62	12.4	162	32.4	86	17.2	190	38	0.022	3.18

Source: *If you look back 5–10 years, how would you evaluate the changes in influence on media content in your news organization from the following factors? (Range your answers on a scale 1–5).* (N =500) The "Journalism in Change" Project Survey, 2012.

A scale of the influence change regarding journalists depends upon the media sector. It is more visible in magazines (+22)[3], advertising-paid free media (+20) and sub-contractors (+39) in contrary to public services media (-10) and paid newspapers (0) (Nygren and Appelberg, 2013:145). On the one hand, these

3 Those who answer "increase" minus those answering "decrease".

studies confirm the general opinion about independent journalism in Sweden, where journalists try to keep the standards created by centuries, but on the other hand, even in the state, which was (and still is) the best pattern of the Democratic Corporatist model of media system:

> "(...) the old "wall" between the newsroom and commercial parts of those companies has been torn down in many ways." (Nygren and Appelberg, 2013:145)

The demands for profit by media owners provokes pressure which has an influence on media content. The border between newsrooms and advertising departments blurs. In the case of Sweden this process takes place but still it is relatively weak in comparison with the two other states. In Russia, it is more visible. Only two Russian journalists fewer than Poland indicate an "increase" (202 vs. 204). In a case of this state we should consider not only influence motivated by economic reasons but also political. Polish journalists express the role of media owner stronger and louder, because they have had lengthy experience in the political system's criticism and they do not feel any fear of publishing negative opinions.

8.4 Conclusion: Consequences of commercialization

The research, survey, and interviews with journalists support the hypothesis that commercialization is in progress in each of the three countries, only the speed and the consequences of this process are different.

In Sweden, the welfare state tradition, journalistic culture (one of the highest in the word) and also a strong non-profit public service sector weaken its tempo. The data confirm that the broad independence and autonomy of Swedish journalists still work, but they have begun to feel more economic pressure around them in the media company. They are aware of an increasing work tempo and the owners' demand for profit, advertisers' influence, freelancers, PR consultants and lobbyists. Foreign ownership is not a problem for them because they are practically non-existent practically in Sweden, apart from a few companies from the Nordic region with the same culture. Finally, the negative consequences of commercialization have arrived in Sweden, but they are slowed by the strong media market, a solid position of owners, a high professional culture of journalists and still high consumption of media by the audience.

In Russia, the free market competition is limited by the structure of ownership. Two main trends are visible there – the media are between etatization and commercialization. Etatization means that the media independence is limited by government on the one hand, but on the other, the state offers some instruments which can weaken a market uncertainty. The television and radio sectors are dominated by the state and state – commercial business ownership. In this context, the opinion of de Smaele can help better to explain the Russian case:

> "(…) media in Russia are not close to politics *or* business but to politics *and* business. Business is politics. (de Smaele, 2010:58)

Russian journalism is located between the state and market, and characterized by the deterioration in the quality of democracy with a decline of media freedom (Vartanova, 2012: 139; Strovski, 2012:308–309). The majority of journalists are satisfied with their jobs (Pasti and Nordenstreng, 2012:247). The Russian media system is a mix of liberalism and authoritarianism, with political and economic instrumentalization. Pasti calls it "the market authoritarianism" (2014), because the logic of commercialization is the same in Russia as Western states and this same tendency towards homogenization of media works, but it happens under the control of an authoritarian regime.

The presented data confirm that commercialization among the three countries is the most advanced in Poland. During the last year we have observed how political pressure was replaced systematically by economic pressure. There are two main sources of this change. The global economic crisis has arrived in Poland. The first reaction of business firms was to cut expenses for advertising. In consequence, it has touched deeply Polish media companies, above all public television and daily national newspaper sector. Many media companies can develop their activity thanks to advertisers and sponsors. Additionally, foreign owners fuel this process. They came to Poland with one aim – to have profit. In general, the Polish market is competitive but some sectors are monopolized by one foreign media group as regional daily newspapers (Verlagsgruppe Passau) or women magazines (Bauer). Polish respondents are the most critical in their opinions about economic changes and the most pessimistic in their vision of future journalism.

Figure 8.5: Relations among elements of media system in Sweden and Poland

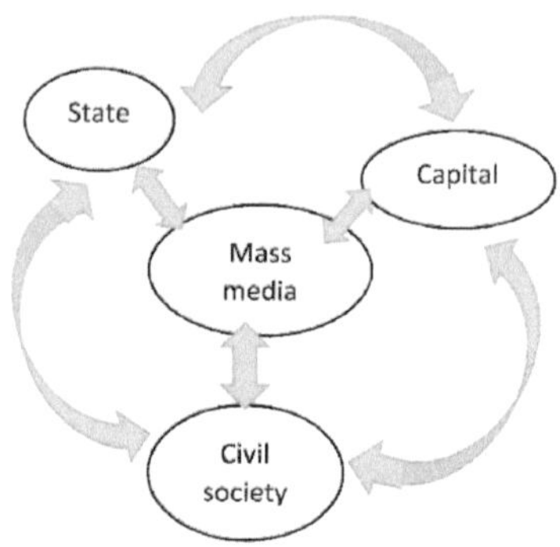

Source: Galtung, 1999:3–21

Figure 8.6: Relations among elements of the media system in Russia

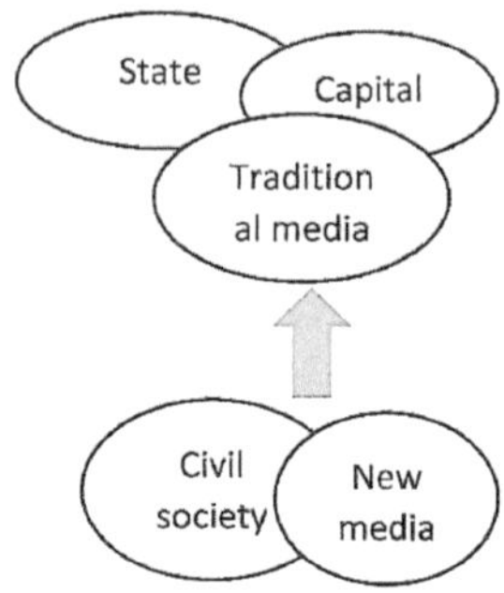

Source: Pasti, 2013.

Our data confirm the different structure of media systems in democratic countries like Sweden and Poland, and in authoritarian Russia. Despite the many deep differences between Poland and Sweden, in a structural aspect the mass media and journalists constitute autonomous organizations strictly separated from political and economic institutions. The mass media act among the political system (state), the media business (capital) and civil society there (see Figure 8.6). In the case of Poland, strong commercial pressure should be emphasized. In Russia, state and private business have very strong connections with traditional media (see Figure 8.7). In fact, traditional media are a part of state-capital structure and additionally, since 2007 state control has been reinforced. Civil society is still weak, however it can use new media such as the internet and social media to express their values and political preferences. But the range of these media is rather limited to the inhabitants of the biggest cities in this huge country and to better educated. The citizens in fact do not have open access to traditional mass media, which serves the government and oligarchs.

Finally, it seems that Poland, more than the other two countries, is touched by the negative consequences of commercialization, fortified by significant foreign ownership, which does not exist on such a high scale in Russia and Sweden. In this aspect, these media systems are in a better situation but generated by different reasons. Swedish owners, editors and journalists developed due to some centuries' old solid standards, high professional values and culture, which helped them to restrain and weaken the negative consequences of commercialization. In the case of Russia, the mainstream media belong to the state and business supported by the state. In effect, real economic competition does not work in the same way as in other democratic pluralist media systems such as Sweden and Poland.

9. New tools for old practices? The journalistic profession in the context of interactive participation[1]

Elena Johansson

9.1 Social media landscapes overview

Digitalization and the emergence of the internet have become the crucial factors challenging the media sector in recent decades. The journalistic profession has become rapidly affected by technological developments, which have led to media convergence in the media environment. It has caused increasing interactivity and opportunities for the individualization of media content. It has also influenced the audience's demands and behavior and consequently challenged the traditional routine of journalists' work, professional practices and even undermined the traditional role of journalists in society.

The first wave of technological changes was the introduction of so called "new media" at the end of the 1990s. Since the middle of the first decade in the 21st century, traditional or "old" and "new" or online media have faced a competitor, social media. "Encyclopedia of Social Media and Politics" describes social media as new forms of media that involve interactive participation (Harvey, 2014:1). An overview of the internet penetration and social media sector in Poland, Russia and Sweden makes it possible to discuss the specifics of Internet use, and some similarities and differences in the popularity of various social media platforms. The penetration of the internet at the end of 2012 was the highest in Sweden and the lowest in Russia, while Poland maintained its position in the middle (see Table 9.1).

1 This chapter has benefited with insightful comments of Jonas Appelberg and Gunnar Nygren, School of Social Sciences, Södertörn University, Sweden.

Table 9.1: Internet, Population, and Facebook Usage in Poland, Russia, and Sweden in December 2012

Country Platform	Poland	Russia	Sweden
Population (N)	38, 415, 284	142, 517, 670	9, 103, 788
Internet penetration (percentage of population)	64.9	47.7	92.7
Facebook users (N)	9, 863, 380	7,963, 400	4,950,160

Source: *Internet and Facebook Usage in Europe* (2013).

The profile-driven global social networking site Facebook was the most popular platform in Poland and Sweden in June 2012. The most popular service in Russia was Russian-language VKontakte (*World Map of Social Networks*). Thus, at the end of 2012 in Poland Facebook reached a share of about 10 million users, in Sweden about 5 million. In Russia, with a population of about 143 million, Facebook amounted to about 8 million users in the same year (see Table 9.1) yielding ground to VKontakte and Odnoklassniki, the Russian-language profile-driven social networking sites. In December 2012 the monthly reach of VKontakte was 29. 5 million people; of Odnoklassniki was 23. 2 million (*Web Index Report*, 2012). Blogging is also popular: Russia is the second top region in the world for LiveJournal, one of the popular blog platforms (Chappel, 2012).

In one sense Poland has the similar peculiarity as Russia: the Polish-language profile-driven social networking site Nasza Klasa is also among the popular services. Since its founding in 2006, Nasza Klasa has been the most visited Polish social media platform. However, by 2010 it was clearly visible that Facebook would overtake Nasza Klasa, and that finally happened in late 2011. In December 2012 Nasza Klasa amounted to about 8 million users. In 2013 in Poland, Facebook took third position (after Google and Allegro) – 73. 6 per cent, while Nasza Klasa only 11th position – 36. 3 percent (*NK.pl straciła...*, 2014).

One more Polish-founded social networking service is micro-blogging platform Blip, the "national answer" to Twitter. However, the closure was announced in 2013mof this micro blog service scheduled for late August in the same year. Users of Blip were encouraged to use Wykop (Allegro Group) and move their contents from Blip to Wykop. The management explained the reasons for the closure as the following: micro blogs in Poland have not become mass utility; strong international competition; a call for new financial investments; the lack of a business model for micro blogging in Poland (Sosnowska, 2013).

The micro-blogging platform Twitter follows Facebook in Sweden; the third most popular is the globally ranked profile-driven social networking site LinkedIn (*World Map of Social Networks*, 2013). At the end of 2012 the penetration of Twitter in Sweden was 10 per cent, in Poland 7 per cent, in Russia – 21 per cent. Twitter is getting up to speed: the number of active Twitter users increased almost two fold in 2012 across the globe (*Twitter Now the Fastest…*, 2013). It would be important to mention that the social media sector is extremely dynamic and changing: new platforms come while the old lose popularity. In the fourth quarter 2013 Instagram has been named as one of the fastest-growing social media services (Lunden, 2014).

9.2 A challenged journalistic profession

Modern-day journalism is changing rapidly in parallel with technological development (Deuze, 2007:141). All news outlets have their traditional platforms as well as being represented online and social media is taking a bigger and bigger place in the work process (e.g. Lewis, 2012:838; Lasorsa et al., 2012:21–23). One big change is that the audience can now participate and produce content. Moreover, the internet and social media has shrunk the distances between all the citizens of the world while relatively cheap smartphones along with other tech gadgets have made production of high quality content a commonality not reserved for professionals. Participatory journalism as a way of collaboration and interaction arises from the phenomenon of participatory culture. Jenkins et al. define it as following:

> "Participatory culture is emerging as the culture absorbs and responds to the explosion of new media technologies that make it possible for average consumers to archive, annotate, appropriate, and recirculate media content in powerful new ways. A focus on expanding access to new technologies carries us only so far if we do not also foster the skills and cultural knowledge necessary to deploy those tools toward our own ends." (2009:8)

The old model of media-audience interaction was based on a top-down structure where journalists had a monopoly to provide citizens with carefully chosen information. In other words it was rooted in the "relatively closed professional culture for the production of knowledge based on a system of editorial control" (Hermida, 2012; 310). Now everyone, at least in the westernized world, could bypass journalists becoming their own editors, producers and gatekeepers, the traits that earlier were monopolized by journalists (Hermida, 2011:178). This rule has definitely changed now. At least this 'new media order' brought pressure upon professional journalists; they should understand what to do about it, what approach to choose and what stand to take (e.g. Allan, 2009:18; Lasorsa et al., 2012: 24; Singer, 2010:117). One of the most important problems for journalists is how to invite the audience for participation without losing control. Newsroom professionals usually see themselves "as kings and queens in the journalism castle"; on the other hand they still try to control the field but on the one hand they more and more perceive now a lack of self-sufficiency and are used to relying on the contribution of others (e.g. Heinonen, 2011:53).

Thus social media challenge the role of the journalist and old journalistic practices and norms. As Singer (2005:180) postulates the popular blog format, as adopted by journalists affiliated with mainstream media outlets, affects long-standing journalistic norms and practice, for example by undermining the traditional role as a non-partisan gatekeeper of information. But at the same time those attributes potentially facilitate accountability as a professional norm. According to her conclusions, journalists (j-bloggers) try to overcome such drastic changes by "normalizing" the new platforms (blogs) to fit the traditional professional practices and norms:

> "(…) most journalists are "normalizing" blogs in at least one key way: they are maintaining control over the information provided under their names, sticking to their traditional gatekeeper function even with a format that is explicitly about participatory communication. As for the related concept of non-partisan presentation of information, these blogs indicate a move away from the neutral stance of the traditional journalist. (…) journalists are taking advantage of the format to strengthen and expand the normative goal of transparency through the extensive use of hyperlinks to source materials. In broader terms, this study supports earlier research indicating that journalists continue to think in terms of their professional role as information providers as they migrate to the interactive online medium." (Singer, 2005:183)

The above mentioned research was connected to blogs but these findings have also been made, and extended, in journalists using Twitter. Lasorsa et al. (2012:30–31) traces the development of the same tendency: journalists who uses Twitter on the one hand "appear to be normalizing micro blogs to fit into their existing norms and practices" but on the other hand "they appear to be adjusting these professional norms and practices to the evolving norms and practices of Twitter". This shows that the normalizing process actually goes both ways. This may be true or not, but social media as Facebook, Twitter and blogs have become irreplaceable tools that journalists use to do traditional journalism now. Sometimes it is easier to reach someone via Facebook then the telephone; Twitter is essentially a large crowd and journalists can easily shout questions and get answers day and night; blogging is a method which could extend the room for professional creativity without any editor in chief (e.g. Knight and Cook, 2013:13; 36–39). A degree of involvement of individual journalists to the process of social media usage for professional needs usually depends on editorial policies regarding social media and can vary at the institutional level. García-Avilés et al. (2013:11) assert that more convergent media organizations hold a stronger and more arranged strategy toward the use of social media by their staff. They develop guidelines for the handling of social media in the newsrooms; encourage individual journalists to use social media in a professional context to spread his/her stories and discussions among the audience and so on. On the contrary, in the less integrated (convergent) media organizations, journalists are not encouraged to use social media and usually do it on an individual basis.

Individual journalists also use social media with different intensity. According to Hedman and Djerf-Pierre (2013:14) there are three different groups of journalists who use social media: the most conservative group being "the sceptical shunners" (sceptics that avoid social media), "the pragmatic conformists" (regular users that use social media primarily professionally and with a certain degree of thoughtfulness) and the "enthusiastic activists" (usually younger journalists, regular round-the-clock/all the week users, privately as well as professionally).

9.3 Research questions

Planning this research we were interested first of all to find to what extent Polish, Russian and Swedish journalists are active in using social media (more exactly Facebook, Twitter, blogs and other social networking sites) for professional goals. Do journalists in different countries use social media for professional work similarly or not? Our preliminary conclusions were that journalists in each country really use social media as tools for professional goals. Moreover, these practices of using them have certain specificities in each country. In other words we found clear differences in social media use by Polish, Russian and Swedish journalists (Appelberg et al., 2014:115).

Now the most important question is what factors can define the differences. Is there a link between the specificity of professional journalistic culture and state of the media on the one hand and use of social media by the journalists on the other hand? What social media platforms do journalists prefer and why? In other words, this research is now focused on a detailed description of each pattern and defining the reasons caused peculiarities of social media use in each country.

9.4 What platforms are preferable and for what purposes?

Our survey confirms that the majority of surveyed journalists in Poland, Russia and Sweden use social media services for professional needs. In the case of Poland a positive answer was within 68 per cent of the sample, almost the same, 67 per cent in Sweden and slightly more in Russia: 76 per cent. Interviewed journalists who avoid social media in their work explain this first of all by fear of being hurt by rude or insulting remarks and comments:

> "As a rule, there are some idiots on social networks that a normal person wouldn't want to talk to. It causes a state of depression and muted horror. (…) ask any journalist and they will tell you that there are always a lot of trolls on any media outlet's website or any social network; people who fill the Internet 24 hours a day." [RR 16, print media, not available data]

> "Social media changed it very much and I must admit that not always it turned to the better. (…) …I thought that in the times of Facebook – where one put his or her name and surname – this manner of hatred will be delimited a little, which is a big enough problem, as it refers not only to us and Poland. As the time went by I noticed that it is such catalyze for such negative emotions and frustration of many people." [PR19, male, up-to 35 years old, public television]

Other journalists are not friendly toward new technologies or find it fussy, in other words the nature of social media contradicts their personality:

> "I use no social media. I think they are unpleasant, unnecessary and I do not understand them. I'm not very social. Others use them, but not me." [SR15, male, 40 years old, reporter, a magazine]

> "I generally avoid social media… On the one hand, they make a journalist's work easier because he/she has access to people and his/her contacts but on the other hand, I appreciate my privacy and e.g. if I had an account on Facebook all my potential interlocutors would know, for example, who my friends are and I don't want people to know who my friends are because I appreciate my privacy." [PR 1, male, under-35 years old, radio reporter, editor, chief of the new media department]

There are also some journalists who actually use social media, but not for work: they just avoid mixing their private life and professional matters:

> "I try not to combine personal and professional things. I never post materials published under my name in social networks. We have a website, groups in social networks, and information contained there is updated by our employees." [RR8, female, 53 years old, editor, Moscow city newspaper]

The majority of the respondents who use social media for professional needs do it of their own free will. However, some journalists recognize that they do it because an editor motivates them: one-third for the Russian sample; for Poland and Sweden the figures are lower. As García-Avilés et al. (2013:11) points out, social media are incorporated into editorial policy first of all in the more convergent media companies. Usually they have a development strategy regarding new platforms and services and encourage their staff to use social media in professional work. At the less convergent (integrated) media organizations journalists usually do it on an individual basis. Interviewed journalists give some examples of such arrangements in the working routine for different media outlets:

> "I personally do not engage in constant interaction with audience… but this channel [TV-channel Russia-24 – EJ] is on Facebook, VKontakte, in Twitter; we are involved in the development of www.vesti.ru [online newspaper of the state media holding VGTRK – EJ]. On the channel, major anchors and key reporters have their own Twitter accounts; they manage a page with our updates. We have special editors who monitor feedback from and interaction with the audience. Normally we try to keep all comments except hostile ones and containing obscene language/" [RR 9, female, 24 years old, State TV, editor of the economic department]

> "We use Twitter and Facebook. It is the main contact through shows, such as in questions in morning shows. Then you can reply on Facebook: for example, this morning, "What do you want to apologize to your siblings for?" It was on the theme of sibling rivalry." [SR 2, female, 39 years old, reporter, radio]

> "(…) the Radio has a Facebook profile and that's where we contact our listeners. Usually our presenters do that and they post information on the Internet, for example, on Facebook about our programs, what our listeners can win, if there are any tickets, what contests we have and so on, the most important information on a given day." [PR 9, female, under 35 years old, commercial radio, reporter, journalist and editor]

However, collective usage does not necessarily mean a high grade of integration of social media into editorial policy. In some cases media management takes a conservative view toward social media in the working process and journalists take the initiative themselves:

> "(…) the chief editor is, unfortunately, a bit anti-internet. There is a website and we have our profile on Facebook but there is a designated person who's supposed to deal with that. (…) The fact that these websites exist, and that we have our profile on Facebook, is because we managed to force the very conservative editor-in-chief to allow us to set them up; but it's only the beginning. I guess he still doesn't understand that there could be feedback." [PR 18, male, over 55 years old, weekly magazine]

Sometimes social media platforms can even work as a compensatory means if media outlets are not represented in the Internet: Facebook or Twitter can take some functions of a media website:

> "We have no blog and no website. And we do not have own website, we think ourselves a problem. Again, this is a decision taken centrally. The company has chosen to work with Play rather than the website. We previously had a home page for our two cities, but not today. I can understand the logic centrally, but locally it will be harder for people who want to have details about how to contact us. We are working with Facebook as "website" instead and from there we get both ideas and are in contact with people watching us." [SR 9, female, 41 years old, TV, reporter]

Global social media platforms Facebook and Twitter are described as "the 'kings' in the realm of social media in the newsrooms" by Garcia de Torres et al. (2011:21) and were the main services used by journalists in different countries in 2011 (Hermida, 2012:315). According to our results the most common platform in all three countries is Facebook (see Table 9.2). Twitter is less popular, especially in Poland. Blogs are extremely popular among Russian respondents, especially for writing. It also seems natural, taking into consideration the data of Social Network Analysis Report (Chappel, 2012) mentioned above. The frequency of using other kinds of communities in the journalistic work is quite high in Poland and Russia and something that can be easily linked with the popularity of local-ranged services: the Polish Nasza Klasa and Russian VKontakte and Odnoklassniki.

Table 9.2: Frequency of social media use for professional work (in per cent of social media users in each age group)

Country	Poland				Russia				Sweden			
Age groups	-35	36–50	51+	All ages	-35	36–50	51+	All ages	-35	36–50	51+	All ages
Read blogs daily	30	35	31	33	61	52	57	59	40	25	21	29
Write personal blog at least weekly	12	13	10	12	44	27	36	40	11	12	8	11
Use Facebook daily	84	62	60	72	75	61	72	72	75	61	46	61
Use Twitter daily	18	26	5	19	35	17	7	30	45	35	17	33
Use other kinds of communities	33	29	34	32	35	15	21	30	21	12	7	14
N=	146	124	45	315	285	84	14	383	97	116	101	314

Source: *How often do you use different types of social media?* The "Journalism in Change" Project Survey, 2012

It is noticeable that the eldest generation of Russian respondents (more than 51 years old) use Facebook and blogs more often than their younger colleagues in the middle age group (36 to 50 years old). A similar difference, putting into question the statement of “more technologically advanced youth” as shown in Table 9.2, can be observed in Poland: the middle generation of Polish respondents uses Twitter more often than the youngest one. The same trend is clearly visible for blogging in the Polish sample: the middle age group is more active than the younger one.

The purposes of using social media in the journalistic workflow vary with social media platforms and countries (see Table 9.3). However, the tendency is that the most important goals in all the samples are: getting ideas and obtaining information (including research, checking facts and investigations) and interaction (a dialog with the audience and contact with colleagues). Advertising or PR by social media is less popular for all three countries and all the chosen platforms. It is remarkable that the respondents in all of the three countries use social media more for contacts with colleagues than to be in dialog with their audience.

There are, however, specifics in using Facebook, blogs and Twitter regarding the purposes. Generally blogs and Facebook are more applicable for information-linked components of professional work (searching of facts, investigation and so on). Facebook also adds a communicative component, for contact with colleagues, as a tool for a dialog with the audience and for self-promotion. It is also connected with discussing social and political questions in general. Blogs are helpful for journalistic investigation mainly for Russian and Swedish respondents, who use it mainly for that purpose. Twitter is an irreplaceable tool first of all for getting ideas and research or checking facts.

Table 9.3: Purposes of social media use (in per cent of social media users)

Country / Platform / Purposes	Poland			Russia			Sweden		
	Facebook	Twitter	Blogs	Facebook	Twitter	Blogs	Facebook	Twitter	Blogs
To get ideas	50	11	38	78	39	66	54	44	62
In research/investigation (to find information)	61	22	37	69	43	71	39	32	44
To have a dialogue with audience	66	9	9	74	35	37	49	33	13
To keep in contact with colleagues	82	7	3	80	31	30	73	33	8
To publish content besides my regular work	18	3	12	61	26	46	23	14	12
To get more readers/viewers, self-promotion	48	8	8	67	33	36	43	35	16
To strengthen the trademark of my media company	49	7	8	58	29	23	37	31	13
For professional discussions	44	10	14	62	21	33	32	30	10
For making money by advertisement or PR	11	0	5	21	9	14	2	0	1
To discuss social and political issues in general	40	9	12	66	26	36	32	22	12
N=*	318			378			333		
* Per cent is based on the number of users of social media for professional purposes in each country.									

Source: *For what purposes do you use different kinds of social media? (Mark yes or no)* The" Journalism in Change" Project Survey, 2012

When putting the countries together, we can see that Russian respondents use every platform more frequently but especially to keep different kinds of discussions: professional and socio-political. Moreover, surveyed journalists also show high results in the publishing of content outside the regular work platform. In comparison to Russians, Swedes and Poles rarely perform any discussion on any social media platform. Swedish and Polish respondents are close to each other in patterns – blogs are mostly used to get ideas and for research as a part of the ordinary work process.

9.5 Conservative professional attitudes in the changed media environment

As mentioned above, interaction is one of the main functions of social media in professional journalistic work. Traditionally, interaction of the audience with the media has always taken place. However, it never was as significant as now, when individuals also have an ability to interact and to collaborate. Generally surveyed journalists have quite positive attitudes towards interactivity. A majority in each country thinks that it is important to listen to the audience and agrees on that increased interactivity influences the selection of news (Table 9.4). However Swedish respondents comprise the largest number of sceptics towards improving the quality of journalism due to increasing interactivity and every fifth says that you should not listen too much to your audience. The journalists in the Polish sample are the most positive but one-third thinks that it takes up too much time.

Table 9.4: Attitudes towards interactivity and journalism

Country	Poland		Russia		Sweden	
	N	% (4–5)	N	% (4–5)	N	% (4–5)
Increased interactivity influences the selection of news	460	60	482	67	454	62
It is important to listen to the opinions of the audience	485	86	492	67	489	65
Increased interactivity can improve the quality of journalism	468	56	486	50	468	38
Contacts with the audience takes up too much working time	475	31	488	32	466	15
A journalist should not listen too much to opinions among the audience	482	15	482	24	490	20

Source: *Interactivity is often discussed in connection to media development. Do you agree or Disagree with the following statements?* (*Give your answers on a scale 1–5*) ("agree"- 4 and "fully agree"- 5). The "Journalism in Change" Project Survey, 2012.

Firstly, there is a clear link between those who think that it is important to listen to the audience and that it improves quality and also that interactivity affects news evaluation (see Table 9.6). Secondly, the ones that think that audience interaction takes up too much time also think that you should not listen too much to the audience. The ones that do not think it is important to listen to the audience do not think that interactivity improves quality either.

Social media have a connective and collaborative nature and thereby make a strong impact on journalistic professional practices, which is traditionally rooted in the "relatively closed professional culture for the production of knowledge based on a system of editorial control" (Hermida, 2012:312). They can even create new forms of journalistic practices. Today's audience's interaction becomes embedded and even interwoven with the traditional media content and the role of social media cannot be overestimated in this process.

According to our results a significant part of respondents in all three countries use social media to be in contact with the audience; however they still prefer traditional ways of interaction such as telephone and e-mails. Generally social media platforms are less popular ways of communication, especially in Sweden (Table 9.5). We can see clear differences in intensity and the choice both between the three countries and between different age groups. The older groups prefer the telephone but even the younger group has more contact through the telephone than through e-mail and social media. The younger generation is friendlier to web-mediated methods, such as social media and comments on the website.

Table 9.5: Using the form of communication once or several times a day (in per cent of all journalists answering the question)

Country	Poland				Russia				Sweden			
Age groups	-35	36–50	51+	All ages	-35	36–50	51+	All ages	-35	36–50	51+	All ages
Telephone	28	34	43	33	25	22	25	24	20	21	35	26
Direct meetings	16	15	13	15	12	11	14	12	7	5	11	8
E-Mail	31	42	49	39	34	31	25	32	27	26	38	31
Comments on the websites	21	28	20	24	25	21	8	23	12	7	4	7
Social media	28	30	21	28	30	22	22	27	17	13	8	12
N=	193	174	81	448	347	117	36	500	123	167	172	462

Source: *How do you communicate with your readers/viewers/listeners?* The "Journalism in Change" Project Survey, 2012.

Summarizing the results, we can conclude that social media are found as an important tool for professional journalistic work by representatives in Poland,

Russia, and Sweden. Social networking is much more attractive for younger users in the journalistic community, which is no surprise and fits global trends. As well as for average users the age of journalists plays a role in the frequency of using social media for professional work. For example, the Russian sample represents mostly younger journalists who, according to Hedman and Djerf-Pierre (2013:15) are usually friendlier toward social media. It particularly correlates with higher frequency of the usage of social media by Russian journalists in comparison with Polish and Swedish. However, the role of age is not crucial. Sometimes the older generation is more active than the middle age generation in using social media in general (see the Russian case in Table 9.5) and in using some social media platforms (see Table 9.2). Generally speaking, such factors as younger age has a lower significance in active social media usage in the Polish and Russian samples. In the Swedish one the difference is clearer: social media are more frequently used by younger journalists.

All three countries are technologically developed and provide high access to social media platforms. However, the results of our survey show fair differences when these new technologies face traditional matters, for example such as professional cultures. This shows that practices of social media use by surveyed journalists in Poland, Russia and Sweden are rather nationally dependent and probably culturally determined matters than a result of technological development.

9.6 Patterns of using social media in professional journalistic work

Thus, when putting the countries together, we can see three different patterns in using social media for professional journalistic work based on specificities of consumption in each sample. Surveyed Polish journalists find important audience feedback. They are less friendly to Twitter but quite active in "other kinds of communities." The latter group of networking services as well as blogging is significantly popular among Russian journalists. They use social media more frequently for political discourse, earning money and self-expression. The Swedish sample is specified by general scepticism regarding increasing interactivity and by some gap between younger and older generations in the practices of social media use.

9.6.1 The Polish pattern

As the results of the survey show Polish respondents are quite active in communication with the audience in both traditional and new ways. Telephone and email contacts are at the top; those figures are much higher in comparison with the Russian and Swedish samples. Contact via social media comes second; Polish journalists do it with the same frequency as the Russians (see Table 9.5). The traditional ways of communication with an audience, the interviewed journalists describe as following:

> "If they [audience – EJ] call me, I always talk to them; if they write, I always reply to their letters (…) and also direct contact, sometimes people talk to me in the street, but I am always surprised that people know what I do." [PR 8, female, 35 – 55 years old, reporter, public TV]

> "We usually keep in touch via e-mails, when they write to me. It's not that we go out and ask the listeners what they think. This is what we have the ratings for, the e-mails, which they write very often. Now there's Twitter, so we can get feedback. There are listeners who are more involved. So I post something on Twitter, something cool, and they comment that it's total nonsense or that it's great. It's a form of communication." [PR 15, male, up to 35 years old, private radio]

However, despite some disadvantage (as for example, unpleasant comments) social media are gradually winning more and more space in communication with readers/listeners/viewers. Moreover, as Polish respondents recognize the importance of this new way of communication:

> "At present it is mainly happening by social media. It has changed a lot because once there were those traditional, common letters, but now it happens very rarely, when someone can switch on a computer and find me on Facebook and send me news." [PR 19, male, up to 35 years, public TV]

> "Obviously it is useful [social media – EJ], it makes contacts easier, but on the other hand it really devours time and my presence on Facebook is, you can say, this is the purely professional presence. (…) I try to be far away from being online all the time on Facebook, because it also means a lot of private messages, a lot of questions. Twitter in this respect is much more secure, you can stay cold a bit… There people do not lead conversations with each other all the time, only they exchange remarks, maybe more essential than on Facebook. It is a more professional journalism instrument when you compare the two big social networks." [PR4, male, 35–55 years old, public radio and weekly magazine]

The vast majority of surveyed Polish journalists (86 per cent) share the point about an importance to listen to the opinions of the audience, which is a record for all the examined groups (Table 9.4). This motivation is clearly explained by the interviewed journalists:

> "(…) nowadays either you work closely with your viewers, your readers and someone identifies themselves with you, or you stop living at all." [PR5, female, 35–55 years old, manager and journalist, public TV]

"(…) there are, of course channels though which they can contact us: they write emails, there is a general Agora [a national newspaper – EJ] e-mail and people really write letters to us, but in the form of emails. We publish some of them because they are really interesting so it is also our readers' influence on the shape of our website or our paper… oh! One more important thing! During our duties for 2 hours on weekdays we are here for our readers and they can come here, just as you came today and report their problems. Recently we started our duties outside the office. For 3 hours I was in Magnolia and readers could come and talk to me … and a few came." [PR2, female, about 30 years old, reporter, a local edition of a national newspaper]

Polish journalists use social media platforms primarily for research and for getting ideas. They also find helpful social networking in contacts with audience and colleagues using mostly Facebook (Table 9.3). Actually Facebook seems a favorite platform for Polish respondents. But according to the interviews each platform has certain advantages and different platforms can be used differently:

"I am on Facebook, but I am not very active there. I don't use Twitter; I don't write a blog… (…) I also have an ambivalent attitude towards bloggers, towards some of them. Because I think it is simply some kind of intellectual masturbation" [PR7, female, 35–50 years old, editor, regional daily]

"I am only on Facebook. Out of necessity, because I also look for people who I might need via Facebook, so I had to create an account. But I am not an active user there… and also direct contact, sometimes people talk to me in the street, but I am always surprised that people know what I do. (…) I use social networking sites in my work but only to find people there. Sometimes I can also find a topic there." [PR8, female, 35 – 55 years old, reporter, public TV]

"There are journalists who have their own blogs and this is an additional form of contact with the reader, sometimes a very heated and interesting one, and sometimes even going beyond the issues touched upon in the magazine. It broadens the spectrum of topics and allows the journalist to express their opinion on subjects that the newsroom rejected but he can write about on his blog, and this is a very positive phenomenon, although a very time-consuming one." [PR18, male, over 55 years old, weekly magazine]

"We make use of Facebook, where we have a profile, or profiles, just in case. We also take advantage of the Russian equivalent of Facebook, which is used mainly in the former Soviet Union; it's called VKontakte.ru… (…) We use Twitter mainly in Poland, because it's a specific kind of medium, a political medium. (…) Polish Facebook is … much politicized, maybe it's because there's no other place where you can express your opinion. I'm convinced that this medium will be constantly expanding…" [PR5, female, 35–55 years old, manager and journalist, public TV channel]

The specificity of the Polish sample is that it has the lowest rate in using Twitter – only 19 per cent (see Table 9.2). It can be explained by the fact of low penetration of Twitter in Poland (*Twitter Now the Fastest…*, 2013) but not only. It would be also reasonable to suppose that local micro-blog service Blip is also attractive for Polish journalists as Twitter. However both reasons could be interrelated. One more noticeable feature of the Polish case is a high usage of "other kinds of communities" (Table 9.2). The answer might be the same: the Polish-language profile-driven social networking site Nasza Klasa is also at the top and still among the most popular services in Poland.

The Polish result regarding positivity towards the importance of audience feedback is higher in comparison with other respondents' groups, especially with the Swedish one. Poles also more frequently suppose that increased interactivity can improve the quality of journalism (Table 9.4). It could be explained by the background of forming Polish journalism. Traditionally it is literature-centric, biased, and politically rooted; it also tends to be quite advocacy-styled (Dobek-Ostrowska, 2012:36; 38). It has also been significantly affected by the almost semi-centennial Communist period. Modern Polish public service media are significantly instrumentalized by the government and political parties, while private media are quite commercialized (Dobek-Ostrowska, 2012: 33–34; 45).

Thus it could be reflected in the much higher respect toward the audience's opinions and in a greater openness to the audience's participation. Polish journalists are probably more open toward an access of readers/viewers/listeners to their professional field. It could be related to the professional identity of Polish journalists: historically it is weaker and less developed in comparison with the Swedish one. One more reason could be linked with the commercialization of Polish media when producing the content becomes consumer-oriented and consequently much dependent on the audiences' demands.

9.6.2 The Russian pattern

Russian respondents use the chosen social media platforms more frequently in comparison with Polish and Swedish. As expected, the number of so-called "pragmatic conformists" and "enthusiastic activists" in terms of Hedman and Djerf-Pierre (2013:14) is higher among younger and middle age generations of the Russian sample. However the same trend is observed for Poland and Sweden. However, as mentioned above, age is not as important, there are examples of older journalists who are friendlier toward social media than younger ones:

> "I highly recommend young journalists to keep in touch with the readers or the viewers through social networks and blogs. Anyone who is engaged in journalism should closely monitor what response is induced by his or her publication, especially in social networks. It is the place where it is much easier to "read" this "audience degree." … I use them frequently, as social networks are now becoming a venue for the news and a place

where you can learn the news that the audience finds the most exciting. I try to monitor social networks every day." [RR 13, female, 57 years old, journalist, national newspaper]

Thus it is not only a matter of a majority of younger representatives in the Russian sample but also a specificity of the Russian public sphere where the Internet plays a role as its substitute. Since the 2000s, and due to technological development, the Russian media sphere became cleaved to "official" and "alternative" (Gorny, 2007; 2011:10). Kiriya (2012:456; 2014:132) identifies the main and the parallel public spheres, where the first one is represented by pro-government widespread TV channels, radio, and some political print media. The second parallel ("alternative") wing is represented by institutionalized (online and "opposite" to authorities) media and non-institutionalized (relatively free blogs and social networking services) media.

Thus, social media, especially blogs, play a crucial role in the development of different forms of socio-political discourses. According to the conclusions of the Berkman centre research, the core of public discourse in Russia maintains in the blogosphere (a constellation of blogs, mainly on the LiveJournal blog platform (Etling et al., 2010:3, 13). LiveJournal is a unique Russian socio-cultural phenomenon: being introduced in 1999 in Russia it quickly became a hit (Gorny, 2011:8; Podshibiakin, 2010:8–10). The modern Russian blogosphere became merged with social networking sites.

What is mentioned above explains such a high popularity of social media among Russian respondents (Table 9.2), especially blogging, which is often associated with LiveJournal. Moreover Russian journalism is traditionally literature-centric (Vartanova, 2012:135–135) and rooted in a "publicist" (writer) role. Russian journalists take rather advocacy function in society by representation of the interests of powerful groups (first of all political). Hanitzsch et al. (2011:283) note that Russian journalists have friendly attitudes toward analytics. All of that is a credit of the blogs: online diary format presupposes development of personification and suitable for writing of long analytical texts. It makes Russian journalists especially friendly toward blogging (Johansson, 2014:35–36; Johansson and Nygren, 2014: 111). Facebook however appears to be the most popular for surveyed Russian journalists: 72 per cent of respondents use it at least once per day or more (see Table 9.2). This is comparable with the Polish and Swedish cases. Frequency of using Twitter by Russian respondents is higher than by Polish and almost similar to Swedish – 30 percent. Russian interviewed journalists describe practices of using those platforms as the following:

"I prefer foreign social networks; the basic one for me is Facebook. There, I also post materials, communicate with the audience, add its representatives as friends. Inadequate comments are extremely rare, I only had to ban two people for all the time I've been active in the social network." [RR18, male, 65 years old, observer and editor-in-chief, web-page niche magazine]

"I may use only Twitter for work; I sometimes tweet about important events I attend. Because the person responsible for the Company's Twitter account is very slow." [RR10, female, 35 years old, correspondent, republic niche newspaper]

Russian respondents pay almost the same attention to usage of other social networking services ("other kinds of communities") as Polish journalists – 30 per cent (Table 9.2). For the Russian case it means first of all extremely popular profile-driven social networking sites VKontakte and Odnoklassniki:

"I don't have a Twitter account, I still don't get what's the perk in it; I don't even know what it is for. I frequently visit VKontakte, I use this network for searching for the people I need…" [RR2, female, 24 years old, editor, state TV]

"I use social networks on a daily basis. Unfortunately, I haven't fully explored Facebook, it's not very user-friendly to me, and I prefer VKontakte. The younger generation is mainly on there, and my colleagues and contemporaries are on Facebook." [RR15, male, middle age group, head of department, national weekly magazine]

Like their Polish and Swedish colleagues, Russian respondents use social media platforms first of all for getting ideas, research/investigation and keeping in contact with the audience and colleagues. Getting readers/viewers/listeners and self-promotion purposes appear to be also in the top five. However in comparison with Poland and Sweden, Russian respondents are extremely active in using social media for publishing content besides the regular work platform, discussion of socio-political issues and using social media for commercial goals as earning money by advertising/public relations and strengthening the trademark of the media company (see Table 9.3).

Regarding the discussion of socio-political issues in the social media, it is not surprising why Russian respondents differ from other participants of the survey. It is particularly true for blogging in comparison with Poland and Sweden. Partially it is explained above by the specificity of the Russian public sphere and popularity of the LiveJournal blog platform that it became a socio-cultural phenomenon in Russia. Thus social media platforms can be used by journalists as a "tribune" but in this sense it is difficult to locate a borderline between using social media for professional and for private purposes. Do they express their political views in the social media as professional journalists or as citizens (Johansson, 2014: 35; Johansson and Nygren, 2014:115)?

The high frequency of publishing content besides regular work can be related to the overcoming of editorial limitations. According to Kiriya and Degtereva

(2010:40) the majority of Russian media institutions (especially TV companies) are under state supervision, directly or indirectly. It impacts on the media and somehow leads to a flourishing of censorship and self-censorship. Social media can thereby work as a compensatory means for overcoming of editorial restrictions (Kachkaeva, 2012:86; Johansson, 2014:36). However this "compensation" can also take other forms. Pasti's (2012:39) describes the typical post-Soviet Russian journalist as "a happy person" who usually combines two jobs: one job in the media company – for stability and the other one (for example as a freelancer) – for satisfaction of creative ambitions. Thus social media have an unlimited potential for the last need. Moreover Russian journalists often use social media for extension of stories, getting feedback, increasing of their own popularity (personal branding) or for overcoming of media format limitation:

> "I prefer LiveJournal, Facebook. I regularly publish all of my finished materials and texts on these social networks. There's also Twitter, but I only have 500 followers on there, and I don't really get it and use it much more rarely." [RR1, male, 35 years old, online media, freelancer, co-works with radio and national newspaper]

> "If the material seems interesting, I post it. I've rarely posted any articles lately… I never directly engage in debates with the audience, namely with writers who post comments." [RR5, male, 50 years old, political observer, local business media organization]

Some Russian journalists are also active in using social media for making money by advertising or PR. The content published besides the regular work might also have a commercial nature and to be profitable. Thus, there is probably some link with commercial usage of social media and publishing content outside regular work: those journalists who publish professionally written content outside regular work can try to derive benefit from social media. In this case social media can be used as an alternative platform for journalists. One interviewed journalist clarifies it by the following:

> "LiveJournal has become a part of information space, and if my materials are good, they are occasionally taken by Lenta.ru, Grani.ru, [online media – EJ] Echo Moskvy [radio station – EJ] Svobodnaya Pressa [newspaper – EJ] I do not focus on an individual editor's office, I have trouble in accepting words like "editorial policy" and "format". A text is good or bad, interesting or uninteresting; the expression "text outside format" drives me crazy. So far I have virtually created my own information channel within Journalism with no Middleman project, money paid by readers can pay for a living if you work and write well. My constant work shows that a journalist can work in a way that makes an editor's office unnecessary." [RR1, male, 35 years old, online media, freelancer, co-works with state radio and national newspaper]

9.6.3 The Swedish pattern

Swedish respondents use social media for professional purposes almost with the same frequency as Polish ones but quite moderately in comparison with Russian journalists. Twitter follows Facebook, and then goes blogs (reading) and finally come 'other kinds of communities' (see Table 9.2) that could probably mean LinkedIn and other global social networking platforms, because there are no local-ranged services in Sweden as in Poland and Russia. Unpopularity of blogging can be related on nationally determined matters: it has never been a common activity for Internet users in Sweden, rather a part of the Internet culture among young Swedish women (*World Internet Project*, 2013:18).

It is also not surprising that younger Swedish journalists are usually more active in using social media: in some ways it reflects the general picture of using social media across the globe. The older respondents feel that younger journalists are in the forefront regarding new technologies; it trails behind what is articulated in the interviews:

> "I feel pretty old-fashioned; I've only got one leg to stand on. I feel like a dinosaur, I can't make websites or web content." [SR 10, female, 55 years old, reporter, magazine]

> "We youngsters grew up in a very different time, with social media, with knowledge of computers, which makes it much easier to work with the changing patterns of communication that the elderly don't, we can use different types of communication and can more easily conquer them." [SR5, male, 32 years old, reporter, TV]

These quotes could be connected to a changed journalistic work process, following the rest of society, where technical literacy is important to be able to function in any given situation. Computer-based methods (social media and website) for contacting the audience have the lowest position in the Swedish sample in comparison with the Polish and Russian ones (see Table 9.5). A reason could be simply that the Swedish sample has more senior representatives than the Russian and Polish groups (see also Chapter 4). Swedes describe a usual communication routine with the audience as the following:

> "First, we have a unique service, a live blog, which we update daily. It works like a chat. We post news and you can ask questions to us. About once a week I sit there – and then I have direct contact with people. Also, I have a lot of e-mail contact. I answer all kinds of questions. Sometimes I answer to Facebook and Twitter in my own name." [SR12, male, 28 years old, web editor, newspaper]

> "I have quite a lot of contact with the readers of Metro [Free newspaper – EJ]. First, when you are out a lot and you're not sitting at the office all the time, but you're out meeting people, then they walk up to you and talk. For example if you do a poll then people want to say something, but above all there's a lot who get involved by calling and emailing. On

> Metro, there is no switchboard so if they call the editorial everyone's phone will ring. (…) On the website, there is a commentary column. (…) I didn't access that and respond directly unless there was a specific question for me. But if people write emails, which happens, I answer, if it is a serious email. I can get maybe 7–8 a day if I have several articles in the newspaper." [SR16, female, 29 years old, reporter and freelancer, newspaper]

Swedish journalists use social media for many reasons, usually different platforms for different things, but getting ideas, investigation and checking facts, contact with colleagues, dialog with the audience, and getting readers/self-promotion are in the top five (Table 9.3). Through the interviews we could get voices explaining this further:

> "I really use social media. If there is someone I am looking for, such as someone who has been through something I can search via Twitter. Sometimes via Facebook if it's something (…) I like to use quite a lot of different forums. For example, I wrote a story for the last spring and I thought that it would be quite interesting to interview someone who did not look forward to the spring, someone who is allergic to the sun. Then I went on a forum for people with allergies and so I wrote a post on the page. Then I get a fair bit of responses if it's a specific forum." [SR3, female, 29 years old, freelance journalist]

> "Twitter has become a very good way to check the news streams, because there are so many of those who publish links via Twitter. (…) I have many different methods for doing this to get the flow sent to me as I want it every day. Twitter is one of many such sources." [SR20, male 42 years old, web editor, radio]

> "I use social media in my work very often. With Facebook you mostly use groups, such as "Missing people" if someone is missing, then you can get tips by following them but also for personal research. (…) But what you absolutely use most is Twitter because there's news agencies and newspapers that first comes with a little message on Twitter before the news telegram because it's shorter to write on Twitter. It simply works as news tips and news mode. You can follow the accounts linked to anything you write. (…) Blogs can be used for research if it is a personal blog. It's very rare that you go into a specific blog without a specific purpose, if so that would be Carl Bildt's [Swedish minister of International affairs – EJ] blog because of his position. Instagram can be used when girls got pictures of themselves posted against their will in Gothenburg. It was based on the Instagram accounts. Then, we used that and also a new chat program called "Kik" that is popular among the kids (…)" [SR19, male, 30 years old, reporter, newspaper]

It is noticeable that talking about using social media for professional needs, the interviewed Swedish journalists mentioned first of all about the collective use of social media (by the media company they work for) and then about individual usage. They did it much more often than for example Russian journalists; this might be a product of the strong tradition of journalistic unions and public service ideals that lines the history of Swedish journalism earlier explained in this anthology. One more feature is that many Swedish interviewed respondents mentioned an unwillingness to mix professional and private matters in the social media:

> "This is something that is still in its infancy. It's a bit of a one way channel. We put things out, there is the opportunity to interact, but it is not often that we get questions there. We see ourselves as news feeds there too. We can comment on news, post more news and talk about our work. If you get a question you answer it. If someone posted, we tweet it. We get more questions on Facebook. There you can go in with comments. But it's very young. I run social media personally, but it is quite another thing if you do it through a company." [SR12, male, 28 years old, web editor, newspaper]

> "It is used for an external track, which belongs to the job. It happens that I sometimes get work related messages on Facebook. I will then try to move it over to my regular mail. Facebook is private for me." [SR14, female, 39 years old, reporter, magazine]

Swedish respondents are quite sceptical toward the statement that increased interactivity can improve the quality of journalism. They think that contact with the audience takes too much time (see Table 9.6). One more peculiarity of the Swedish pattern is that using social media for maintaining contact with colleagues is even more important than maintaining contact with the audience, which is especially visible for Facebook (see Table 9.3). These specificities can be related with to the background of Swedish journalism. Using Hanitzsch et al. (2011:280) terms and described in Chapter 2 of this anthology, Swedish journalists represent a "Western journalism culture" cluster, it means that they traditionally follow ideals of detachment and consider themselves as watchdogs and a fourth estate. Moreover Swedish journalism is strongly rooted in public service values.

Historically the Swedish journalist has had the mandate to monitor society and to set the agenda following their professional norms and ideals. The striving to control the field can be a part of a strong professional ideology, a typical situation in the media milieu where strong gatekeeper ideals and public service traditions flourish. Thus, such an "elitist" position of Swedish journalists in society makes difficult to invite the audience into the work process. The senior generation has an over representation in the Swedish sample; usually older people have a more conservative outlook and hinder the development of some new practices.

9.7 Conclusions and discussion

It has been aimed in this chapter to find some similarities and differences in using social media for professional needs by Polish, Russian and Swedish respondents and to clarify the reasons that can define such users' behavior. Indeed the majority of surveyed and interviewed journalists recognized that in providing the working process, they are really armed with social media services. However, despite the high frequency of use, social media has not replaced older communicative means as e-mail or telephone. It shows the use of these platforms as a compliment to the more traditional ways of communication, not changing the work process, tending to "normalize" or to fit the new tools to the old practices as explained by Singer (2005:174).

This case study shows that the age criteria are not crucial in the practices of using social media: sometimes the older generation of journalists can be more active than the younger one. The technological factor is also not so important; internet penetration can be relatively lower, as in Russia, but it doesn't lower the frequency of using social media for working needs by the surveyed journalists. Moreover, there are examples when interviewed journalists avoid social media, even while working at technologically developed companies with incorporated social media policies. Thus the using of social networking is also a matter of journalists' personalities.

Garcia de Torres et al. (2011:21) identified Facebook and Twitter as the most frequently used platforms by journalists in many other countries and determined these platforms as "the newsrooms social media 'kings.'" Indeed Polish, Russian and Swedish respondents name Facebook as the most preferable and favorite platform: it follows the global trend. However there are also certain differences regarding the frequency of Twitter usage: Polish respondents are not as active in Twitter as Russian and Swedish ones. There is also some difference in relation to other platforms: Russian journalists are extremely active in blogging. Together with Polish respondents they also find quite helpful the "other kinds of communities."

In the top purposes of usage are getting ideas, research of information and keeping in contact with the audience and colleagues. This is true for all the groups of respondents. It clearly shows that use of social media among journalists is "normalized" into old journalistic processes, where these functions traditionally play extremely important roles. In this way social media extend the 'tool kit' for providing the familiar working routine. But on the other hand, we can also see differences. The biggest gap between the samples lies in the use of social media for socio-political discussions, publishing content outside the regular workplace and for earning money: Russian respondents do this much more often than Swedish and Polish ones.

Talking about differences in use of social media in this case study we can identify the following reasons: technological, political and socio-cultural; the last two are usually closely linked with peculiarities of professional journalistic cultures. The ranges of differences shape the patterns of social media use. In such a way extremes in the choices of social media platforms might simply be a matter of access to more convenient services. This convenience can be both technical (immediate access, availability) and cultural (local language, local traditions/socio-cultural practices). Thus, Polish and Russian respondents use 'other kinds of communities' much more frequently than Swedish. This is because they have an alternative: popular local social networking sites as Nasza Klasa in Poland and VKontakte and Odnoklassniki in Russia. Probably Polish journalists were also less active in Twitter because they had an alternative local-range micro blogging platform, Blip.

An influence of socio-cultural heritage and political situation plays also a significant role in the forming of the above-mentioned patterns. Blogging has a literary nature and is close to classical journalistic genres. On the one hand literature-centric traditions of Russian journalism in combination with the specificity of the public sphere can be seen in the wider use of blogs among respondents. On the other hand strong political pressure and a lack of press freedom made the blogosphere in Russia an important space for public discourse and for the publishing of uncensored content. In Poland and Sweden blogs play a minor role for journalists. In Russia, social networking can also bring almost unrestricted opportunities for professional self-expression and creativity of journalists, especially for low-paid employees at the state/state-controlled media organizations.

Strong traditional views of professional culture might lead to active resistance to the levelling effects of social media: the majority of Swedish respondents demonstrate quite sceptical attitudes toward increasing interactivity. They prefer to keep the audience out of the professional field, and probably this is based on the developed professional identity and elitist position of Swedish journalists. It goes along with using social media rather for contact with colleagues than for contact with the public: professional discussions in social media have the same or even higher significance as a dialog with the audience. But in practice it is also true for Polish respondents despite their positivity toward audience participation. Could it display some factual disinterestedness of all the surveyed journalists in audience participation, which probably does not depend on the differences in professional cultures? Thus, if it would it be possible to find a link between specificities of professional journalistic cultures (whether it is arise from strong principles of journalists' detachment and gate-keeping function as in Sweden or from being developed from literature-centric and "biased" heritage with the apparent advocacy role of the journalist in society as in Russia and Poland) which somehow affects the patterns of social media use, can we also speak about some general and universal basis which works similarly for all of the journalistic communities?

It is important to remember that the patterns of using social media are obviously based on complex different factors. What we can see might not be that journalists fit into the participatory culture in that they have a dialog and cooperate with the users. Thus the main reasons for using social media stated by the majority of respondents are to get ideas, not keeping a dialog with the audience. Sometimes even dialogs with colleagues have a higher priority. This is in line with the earlier research of normalization, which states that new platforms or tools do not necessarily change traditional journalistic ideals and values. Instead they are integrated into a traditional work process becoming new tools for traditional tasks. The journalists engaged in an increasing indirect dialog with the audience and used them for minor crowd sourcing, listening in on their conversations and taking part of their privately produced content. This is nothing new, but since the boom of social media and mobile platforms audience interaction and participation in overall societal conversation has become much more visible and therefore reachable for journalists.

A weak spot of this case study is a lack of monitoring of the role of social media in the journalistic working process, which one could conduct both on individual and institutional levels One more dimension could be an analysis of the audience's participation. The conclusions above are based mostly on the respondents' answers regarding the practices of using social media but not on the facts, which reflect reality. Thus it might be a topic for future research. This paper, however, puts under the spotlights the practices of using social media in the working routines by professional journalists in Poland, Russia and Sweden.

10. Journalistic cultures between national traditions and global trends

Gunnar Nygren and
Bogusława Dobek-Ostrowska

To be a journalist in Poland, Russia and Sweden means – in many ways – to be a part of the same community: the most important ideals are the same, the daily work is performed with similar tools, formats and expressions are similar. On the surface there are many similarities, but still there are important differences when it comes to the conditions for professionals. The frames in the daily work describing the boundaries for each journalist are different – commercial and political limitations for what is possible in journalistic practice. There are commercial pressures from owners and advertisers and political influence on media companies and newsrooms on journalists to adapt in their daily work. Additionally, what is more important, these limitations differ between the three countries and among different types of media.

These are some of the results presented in previous chapters in this anthology. The point of departure for the project has been the rapid changes in media technology, society (changes in users' behaviour) in combination with crisis for business models and increasing market influence in the media sector. The question is how these changes influence professional journalistic cultures in different media systems. Poland, Russia and Sweden represent different traditions in journalism, and the position of media in relation to political power and society differs as well. We knew from the beginning that there were differences – but is it also possible to identify similarities between journalistic cultures in the era of globalization of professional cultures (Waisbord, 2013)?

The project has researched the changes in journalism from the perspective of representatives of this profession. With a survey to a representative sample of 500 journalists in each country, opinions and experiences were studied. In 20 in-depth interviews in each country journalists were asked to give more detailed answers going beyond the questions in the survey. All the empirical work was accomplished by national teams in 2012, and the results were analyzed and presented at international and national conferences in 2013–2014 (see Chapter 2).

The results of the project have been presented in thematic chapters. In this final conclusion we summarize and present answers for the following research questions:

- How is technological and economic development influencing professional journalistic cultures in the three countries?
- Do the changing practices influence journalists' perception of routines and values?
- What are the consequences for professional autonomy – in daily work and in relation to other social fields such as politics and the economy?
- Is media development making journalism more similar in spite of differences in traditions and media systems? Is there a homogenization of journalism or is it more correct to label it hybridization?
- Is the status of journalism as a social field changing; is there a de-professionalization as a result of media development?

10.1 Results in three dimensions

Cultures have different dimensions, relating both to material basics, practices and patterns in thinking. Professional journalistic cultures have earlier been described in three dimensions: 1) a cognitive level about how journalists shape the world, 2) ideals with beliefs and values and 3) the role of journalists and practices in the daily work (Hanitzsch, 2007). In our analysis, we also define three dimensions but we describe them slightly differently. Although two are the same: ideals and practices, the third dimension was defined as the conditions for journalists to form a coherent social group. This includes the degree of professional education, the existence of professional institutions, if it is possible to make a living from journalism and the mobility in and out of the profession. These material conditions form a basic dimension of the professional culture, if they are weak it also influences the other dimensions of practices and ideals.

An analysis of the result from the project in these dimensions gives this general picture of the cultures (see Table 10.1):

Table 10.1: Three dimensions of journalistic cultures

	Poland	Russia	Sweden
Journalists as a social group:			
– Professional education	Medium	High	High
– Professional institutions	Low	Low	High
– Share of income from journalism	Low	Medium	High
– Mobility in/out of the profession	Medium	High	Low
Journalistic ideals:			
– Neutral reporter	High	High	High
– Activist educator	High	Low	Low
– Advocate of justice	High	Low	High
– Degree of detachment	Medium	Low	High
Journalistic practice:			
– Multiskilling	High	Medium	High
– Autonomy in daily work	High	High	High
– Political influence in work	Medium	High	Low
– Commercial pressure	High	Medium	High

Source: Authors, based on findings from "Journalism in Change" 2014.

10.1.1 Journalists as a social group

In Poland, the journalistic community is rather weak. Nearly half of the respondents lack professional education, and only about 30 percent belong to professional and trade unions. Half of the Polish journalists surveyed say they can only survive on their income from journalism or they have to get other incomes. Still, most Polish journalists want to stay in the profession, although a majority believe the profession will vanish or transform to something else in the future.

In Russia, the pattern mirrors that of Poland with some exceptions. Only a small minority belong to professional organizations, and many journalists need incomes from other activities. A large share of respondents hold journalism education, but still 25 percent of them think they will be in another profession within five years. Journalism is considered to be a "social lift," according to Russian researchers (Pasti et al., 2012:279). Part of this can be explained by age – about 70 percent of the Russian journalists in the survey are under 35, and at the beginning of their careers. The soft border around the profession offers a high degree of mobility in and out of the profession for Russian journalists.

In Sweden, the situation is different compared to the two post-communist countries. Journalists form a quite stable social group, according to the results. Nearly all of the journalists surveyed have a professional education, and about 90 percent of them belong to professional organizations. Only a small minority needs other incomes, and journalists participating in our research want to stay in the profession – unless they are going to be retired (13 percent). One explanation can be the aging group of journalists in Sweden, with nearly 40 percent over 50 years of age. Among young journalists in Sweden the tendencies are the same as in Poland and Russia with a lower degree of membership in professional unions, more temporary jobs and a larger mobility in and out of the profession.

All together, the analysis shows both differences and similarities for journalists as a social group. In Poland and Russia, journalists have become a fragile and liquid profession with few institutions and social bonds keeping the profession together. The Swedish journalists form a much more stable social group in general, but from a generational perspective there are clear similarities between young journalists in all three countries: Journalism becomes a more fluid profession with a greater degree of mobility in and out and professional structures are becoming gradually weaker (a development also described in other countries by Deuze 2007:236).

10.1.2 Journalistic ideals

When it comes to journalistic ideals, the results show both important similarities and differences. In the analyses, three dimensions in ideals were found in a factor analysis (see Chapter 6). A comparison of these dimensions plus the degree of detachment in relation to other fields gives a picture of how journalists perceive their roles.

In Poland, the most important ideal is to be a neutral reporter. But also more active ideals as being an advocate of justice and an educator scores highly among Polish journalists. A clear group (about 20 per cent) feels no problems also working in PR or to taking gifts from other sources of information. But in relation to politics, the borders are clearer: the interest in political issues is high, but nearly no one thinks it is ok for journalists to be active in politics. The borders in relation to PR seem to be much softer than in relation to politics; political influence in media companies is not tolerated among Polish respondents.

In Russia, the ideal of being a neutral reporter also is the most important. Other more activist ideals are weaker; the ideal of being an advocate of justice is less important. Russian journalists have no tradition of controlling public authorities and being a public watchdog, more important are passive ideals of objectivity and bringing forward various opinions. This is a clear change, compared with earlier studies from the 1990s, where Russian journalists expressed more activist ideals (Wu et al., 1996:544, Pasti and Pietiläinen, 2008:113). The new "Putin generation" of Russian journalists seems to have more passive and neutral ideals and at the same time the borders towards other areas are quite soft for many Russian journalists. About 30 percent of respondents think there's nothing wrong with working for the PR industry and about every second Russian journalist can imagine themselves working in politics. Only four out of ten Russian journalists surveyed say no to mixing a political career with one in the media.

Also in Sweden, the ideal of neutrality is at the top. Nearly at the same level is the ideal of being an advocate of justice. To be a watchdog is important for Swedish journalists, also in other research (Wiik, 2012:34). To support this, Swedish journalists surveyed tend to draw very clear borders for integrity of the profession: very few think it is ok to work with PR, and nearly nobody is willing to receive gifts from sources. The interest in political issues is large among Swedish journalists, and also a substantial group think it is ok to be active in politics. But nobody can accept political influence in the media company.

To summarize: The most important dimension of ideals is the same in all three countries – to be a neutral reporter. This ideal is closely connected to the classic liberal ideology of journalism as an independent but passive spectator. More active ideals differ between the countries: the watchdog-dimension is strong in Sweden and Poland, but not in Russia. The educator-dimension is quite strong in Poland but not in Russia and Sweden. This difference between passive and active ideals also relates to the degree of detachment among professionals – with active ideals of being watchdogs and educators, journalists are also more eager to keep clear borders towards other areas as PR and politics.

Earlier research has stressed the plurality of professional role perception, mostly journalists combine different roles in their own professional identity (McQuail, 2010:285). This is also the case among journalists in Poland, Russia and Sweden; the interviews give many examples of this mix. But there are also cultural differences and different patterns in Sweden and Poland compared with Russia. The subjective tradition of European journalism has combined activist ideals with close ties with politics, and the public service tradition has combined objectivity and neutrality with a detached position in relation to politics (Donsbach, 2010:41). This is also the pattern in Poland and Sweden where there is a clear line between journalism and politics.

In Russia, the more passive/neutral ideals among journalists are combined with softer borders around the profession in relation to both politics and PR/communication. One explanation for this can be that Russian journalists use the neutral ideal as a defence towards political pressure. This attitude is also visible in interviews with Russian journalists concerning gender aspects in reporting: the reporters refuse to consider any other aspects than objectivity and any kind of gender sensitivity is perceived as a potential threat to their freedom of speech (Voronova 2014:112). Another explanation can be that Russian journalists turn towards a commercial tradition (Donsbach, 2010:41) with a more passive and subordinated role for journalists to produce what the company wants for economic success (within the political limits of the system).

10.1.3 Journalistic practices

There are many similarities when it comes to journalistic practices in the three countries. Questions from the survey indicated many similarities with regard to the impact of new technologies and changes of daily work in the newsrooms. In general, the respondents describe these processes in a similar way. In their opinion, new technology has the same impact in daily work. But we cannot forget about a significant difference in the development and application of these technologies in each of the three countries.

In Poland multiskilling is frequent, and the attitudes are positive among journalists. Multiskilling is understood as both covering different areas and working for different platforms and with the whole process (research, photo, producing/writing, editing). Polish journalists think this increases creativity and power in the daily work. Polish journalists surveyed also perceive a large degree of autonomy in decisions about subjects and emphasis in the stories. But they feel a very strong commercialization in the daily work, and the strong influence of advertisers and owners. This commercial influence limits their autonomy more than political influence, which seems to be lower than in the past.

In Russia, multiskilling is more common in certain kinds of media like TV and online journalism. In newspapers and in large media organizations, traditional professional roles seem to be more common. Russian journalists report a high degree of autonomy in daily decisions, but there is also clear political pressure and limits for what is possible. The political pressure from owners and from political actors outside the media company is strong and nearly half of Russian journalists report about decreasing press freedom. In the interviews, Russian journalists give many examples of a dual experience; they perceive a high degree of autonomy in selecting stories and angles as long as they keep within the limits of editorial policy. These limits are both political and economic, and create a high degree of self-censorship among Russian journalists – by experience they learn to adapt and to act within given limits. Another part of the Russian experience is to choose a place of work from political perspectives, to work in a media company similar to your own political attitudes (Nygren and Degetereva 2012).

In Sweden, multiskilling is used among journalists in all kind of media. The demands on increasing production are very clear, and journalists perceive this as a threat to their professionalism. They have a high level of autonomy in daily decisions, but they also report an increasing influence of commercial parts of the media company both to increase production and in strategic decisions about journalism. Political influence in journalism is very low, almost non-existent. Problems mentioned by journalists are downsizing of newsrooms and demands on increased production in cross platform publishing. In the interviews, many Swedish journalists mention economic discourse being dominant in the media companies, it is not about journalism anymore.

In general, we can conclude that there is a move towards a larger degree of multiskilling, even if the speed of development differs between media types and countries. It seems clear that all journalists perceive a certain degree of autonomy in the daily decisions on choice of subjects and angles, but also differences when it comes to the limits for this autonomy. There are different kinds of external pressure on the newsroom – in Poland and Sweden mostly commercial pressure and in Russia political pressure from owners and external political actors is the most important obstacle to professional autonomy.

10.2 Comparative journalism research

It is impossible to describe a culture in a few paragraphs. Every attempt will meet the risk of sweeping generalizations, too much simplification and to 'avoid the complexity of reality. With these remarks, there are still some patterns in the journalistic professional cultures in Poland, Russia and Sweden to describe as well as common patterns for journalistic cultures in all three countries.

In Poland, the journalistic profession seems to be weak on a collective level. At the same time, the ideals and the ambition to influence society is strong among individual journalists. The memory of the communist system and also how politicians have influenced journalism after the fall of communism give Polish journalists a strong sense of integrity in relation to politics. But it is harder to resist commercial influence in the media companies, interests of advertisers and owners influencing professional decisions. Foreign owners have no political agenda known to Polish journalists, but commercial demands put pressure on journalistic work.

In Russia, the tradition of journalism is strong with literary roots in the 19th century, and a large share of journalists has a professional education. But at the same time the borders around the profession are soft, and mobility is high in and out of the profession. A large majority of journalists is young, about 70 per cent under 35 years of age and many of them perceive journalism as a step on the career ladder to something else. The ideals are mostly to be neutral and passive, and journalists have to adapt to pressure from other actors. This pressure is very much political, both from owners and political actors. At the same time, individual journalists seeks their own spaces on the net, in blogs and social media to express their own opinions and discuss social and political issues. In the blogs, professional roles of journalists are mixed with other roles of activists, of the private sphere etc.

In Sweden, the profession is strong and coherent, the borders in relation to other areas well defined and integrity high among journalists. The ideal is very much a "detached watchdog" (Hanitzsch et al., 2012:485), and to fulfil this ideal journalists have the support of strong professional institutions. But at the same time the professionals are old, nearly 40 per cent are over 50 years of age and there is sluggishness when it comes to change. The attitudes towards multiskilling are critical, and journalists are critical towards increasing commercial influence. The professional culture is under pressure of downsizing in newsrooms, of economic strategic thinking and many journalists in the interviews question their future in the profession.

These patterns show differences. But there is also a common pattern of a professional culture transcending national borders and cultures. It is possible to identify some parts of this common professional culture in the results – professional education for a majority of the journalists, neutrality and objectivity as the most important ideals, quite similar processes in daily work and a certain degree of autonomy in the daily decisions on subjects and emphasis in the stories. To be a journalist is in this sense the same profession – but it is also rooted in national culture and differs when it comes to political and commercial conditions.

Ideals and reality

These similarities are also found in earlier comparative research (Deuze, 2007:163). In a survey of journalists in 18 countries, Hanitzsch et al. label these similarities as a "general cultural understanding that is shared by most journalists around the world" (2010:15). Other researchers emphasize the origin of these ideals from the classic liberal model, and describe a convergence between media systems towards the liberal model. Even if important differences still persist (Hallin and Mancini, 2004:301).

Differences have been analyzed in relation to cultural dimensions, often focused on institutional roles of journalism. These roles also explain some of the results in the project "Journalism in Change" – the degree of interventionism (passive or active dimension), the distance to political power and the degree of market orientation (Hanitzsch, 2007:371). The results show clear differences in these roles in Poland, Russia and Sweden, in the perception among journalists about how a professional is supposed to think and act.

But there are also clear differences in the position of the profession; in the strength of professional institutions, if you can manage on the income from journalism and the conditions in work in relation to owners and power. All these basic material conditions also matter when it comes to final professional performance. The question is to what extent journalists are able to fulfil ideals. This is what Weaver and Willnat discuss in "The Global Journalist" (2012:544) – there are many similarities in ideals but also differences reflecting societal differences in culture and political systems more than influences of media organizations and professional norms.

This leads to the question if we see a homogenization of journalistic cultures, a kind of convergence around the Western liberal model? The results give some support for this hypothesis, for example the fact that the "neutral reporter" is the strongest dimension in role perceptions in the three countries. But looking at other areas, the development is rather to be labelled as a hybridization of journalistic cultures. New technology and changing processes are adapted to the old familiar culture, for example how Russian journalists take blogs and social media as a new space for the alternative "kitchen-table" discussions. Important values as being a "watchdog" on power are very important to journalists in Sweden, but not in Russia – and this is not changed by media development. In Poland the autonomy in relation to the political sphere is crucial for journalists, probably as a result of experiences from the past. This influences Polish culture strongly, in contrast to the Swedish situation.

There is an impact from globalization and technological development on journalism in all three countries. This impact is taken into the professional culture, and transformed in an on-going negotiation with history and traditions of journalism in each country. Differences also arrive from differences in organizational cultures; each organizational culture is unique in relation to values, norms, behaviours, the role of managers/leaders, heroes, myths and stories developed and cultivated in a given media enterprise.

10.3 Journalistic cultures and media development

Digital technology has changed the media landscape globally. New tools and productions processes are the same all over the world. The Internet has changed the daily work for all journalists, both as a means of communication with the audience and sources and as a platform for publishing. Social media creates new public spheres both for citizens, journalists and media companies.

All this is clearly visible in the survey and in the interviews. The old journalists in interviews talk about all the changes, how the daily work has changed. At the same time basics in the daily work remain the same, regardless of changes in technology: to find and evaluate news, to reach sources to verify and to publish for the audience. The tools and speed of the processes has changed everywhere, but the basic steps in the process remain the same in daily journalistic work.

Still, media developments give professional culture some fundamental challenges (see also Chapter 1): questions of professional control or open participation, of journalism as products or processes and the restructuring of the media industry. These challenges change the professional culture in different ways:

Professional control or open participation

There are no signs of journalists giving up their professional control over the production process, at least in the eyes of journalists themselves. New tools and means for communication are normalized into the old practices, for example social media becomes an important tool to reach both old and new sources but it does not change the control over the production process. Communication with audiences becomes more important, but still the new tools are mostly normalized to do the old things.

It is possible that new online networks provided by social media can strengthen the role of the individual reporter or at least the personal profile. Journalists are no longer dependent on the printing press or broadcaster to reach an audience, and this becomes important in countries with limited press freedom like Russia. The survey shows that about 40 percent of those Russian journalists using social media regularly use blogs for publishing of content beside their regular workplace. The Internet and social media can also give journalists access to new kinds of sources, for example by using methods in data journalism and crowdsourcing. As conductor of the process, journalists are still in control of it. By using new technology, the position of journalists can even be stronger in relation to both traditional sources and in relation to media owners.

In a larger perspective, professional journalistic control over communication processes in society has disappeared in the interactive networks of the Internet.

Sources have new ways to communicate with audiences, and are not forced to pass professional "gatekeepers" in traditional media. All kinds of users build their own networks in social media and other online-based media formats. All this gives both sources and audiences new patterns of communication.

This also influences journalism. Open participation challenges the professional control and the old one-way communication (Lewis 2012). In the network society, journalism is reduced to one of many forms of communication. But in this reduced journalism news is still produced by professionals according to professional standards and with professional control. The challenge for journalists is to integrate voices and viewpoints of others within the networks to a greater extent:

> "Journalists will never again control the flow of information in the way they once did... Journalists in a network must acknowledge that they will retain power only to the extent that they share it; without facilitating the broad exchange and not merely the delivery of information, they will find themselves increasingly irrelevant to the conversation taking place around them." (Quandt and Singer, 2009:141)

Journalism as a product or a process

Cross platform publishing is developing in all three countries, but at different speeds. Media companies publish in different channels, and there is a convergence between genres and formats in different kinds of formats and channels. This creates a new kind of polarization among journalists, between a large group of young, multi-skilled "production journalists" and a few specialized profiles. At least in Sweden, journalists more seldom leave the newsroom, instead they work at the desk. Recycling of content and adapting raw material to different formats becomes more frequent. But there are also other journalists, specialized "stars" and columnists giving each media company their profile.

Publishing 24 hours 7 days per week, changes journalism to be more like an on-going process. For some older journalists mostly in printed newspapers, it is difficult to change from writing an article to publishing small pieces many times on the web. But in the long run this can change the professional culture from producing individual products, to journalists being more like participants in a process. In online journalism this is visible when the journalist no longer "owns" the story, but makes his or her contributions to the story during the working hours. The process becomes larger than the individual journalist.

Restructuring of the media industry

This is perhaps the most difficult challenge for the professional culture of journalists. The changes are most visible among young journalists: many have temporary employment and there is a large mobility in and out of the profession, some combine work in journalism and other areas in media like PR, communication or creative occupations. Working life becomes more fluid, and the borders of professions less strict (see also Deuze, 2007:154). New business models make journalists more responsible for business results, commercial values become more important.

All this also influences professional identity, it is likely to change from being a journalist to more like a "media worker" at least among young professionals. The old kind of journalists working on a steady job in a media company until they retire is becoming rarer.

The trends are the same in all three countries, but the speed is different. In Poland, the changes towards a more fluid labour market seem to be at a high speed. In Russia, the labour market is more stable but instead journalists show more signs of mobility in and out of journalism. In Sweden, the generational gap is significant with an old stable part of the profession, and young journalists struggling to find some place to work.

Depending on the point of departure, the changes in media industry and in production have different kinds of meaning for journalists. Multiskilling can give the individual journalist more power in the work, according to opinions among Polish and Russian journalists. One explanation can be these journalists having a limited influence before, and now with access to new tools they can control the whole process. In Sweden the attitudes are the opposite, and the explanation can be that Swedish journalists already had a lot of influence and now also have to do work earlier done by photographers and editors. Another example is the different meanings of temporary employment – if there are a lot of jobs on the market, a temporary job is not a problem for the individual journalist. But if there is a lack of jobs, journalists employed on a temporary basis become more dependent on the employer. In the survey, it is clear that the temporarily employed perceive less professional autonomy in the daily work, they have to adapt more to the newsroom and the editors.

Identities and cultures are never stable: they change over time due to external influences and internal processes. This is also valid for professional journalistic cultures, and the results point in the direction of a weaker and more fluid professional identity. It will be more difficult to define a special common professional journalistic culture in the growing convergent media culture (Deuze, 2007:235). But still, individual journalists might gain from the changes, use new technology and develop new forms of journalism and a new kind of professional autonomy in this changing media landscape.

The development will also differ between media systems – the consequences for professional journalistic cultures depends on from what history you start the journey into the new media landscape; what is a threat to journalism in one country can be a new possibility in another.

10.4 Journalism as a social field

Journalism is a weak autonomous field, according to Bourdieu (2005:5). It has a position as mediator between other fields as politics, economics and culture, and all these fields are intertwined in the field of journalism. Still, it is possible for journalism to gain some degree of autonomy as a social field by developing a specific cultural capital. Social networks, such as professional associations and symbolic capital as values and ideas can give journalism some kind of autonomy (see also Chapter 1).

To analyze journalism as a social field is to think relationally, to analyze the degree of autonomy for journalism as a social field and the autonomy for journalists in the daily work. In the project "Journalism in Change," important questions have been how political and economic spheres influence professional culture. How journalists perceive their autonomy in daily work and the influence of external and internal political and economic actors are key questions, both when it comes to a general influence and special parts of the process as choice of subjects and as obstacles for publishing. From the results on these questions, it is possible to discuss the position of journalism as a social field.

There are clear differences between the countries in the relations between journalism and other social fields.

In Poland, journalists report a decreasing political influence and increasing press freedom. They are very clear in drawing a line in relation to politics – no mixture is accepted among Polish journalists. But in relation to the economic field, the borders are more blurred, and Polish journalists report an increasing commercial influence in journalism, which was one of the topics analyzed by the European MediaAct project (Dobek-Ostrowska, Głowacki, Kuś, 2014: 156–158). There is a symbolic capital in journalism in the long tradition of resistance against political power, but still the field has weak institutions and a low degree of social capital. High mobility and bad economic conditions make journalism vulnerable.

In Russia, journalism can be considered to be a weak social field in many ways. Journalists report a strong political influence on all levels and a decreasing press freedom. Journalists perceive some degree of autonomy in daily decisions, but at the same time the limits of what is politically possible are clear. At the same time cultural capital of the profession is weak, mobility to other areas is large and the sense of common professional journalistic values like ethics less developed than in other countries. Russian journalists are part of the traditional "intelligentsia" with close connections to political power and the cultural field as literature, a field larger than journalism and with less clear professional borders within (Trakhtenberg, 2007:126–127).

In Sweden, the political field still gives some frames for journalism in maintaining a strong public service in radio and television and some subsidies for smaller daily newspapers. But in general, Swedish journalists perceive a low political influence in professional journalism, in spite of the tradition of party press and the system of state subsidies to small newspapers. The economical field is much stronger within media companies where commercial pressure is increasing. This influences both conditions for producing quality journalism and strategic decisions in the newsrooms. At the same time, journalism has a strong cultural capital in strong professional institutions and a strong public service orientation. The balance between the fields is moving in favour of the economy, but media companies also use the symbolic capital of journalism as a commercial asset in competition. Journalism still has a great deal of autonomy, as long as it does not come into conflict with commercial interests of the media company. At the same time, as the professional borders around journalism are getting softer, it becomes more difficult to define a journalist in relation to other social fields.

A position rooted in history and culture

The analysis shows that the position of journalism as a social field is deeply rooted in history and national culture of the three countries. In Russia, media researchers have shown how journalism has developed under the authoritarian regimes as a tool for power to inform society, and at the same time for elites (the "intelligentsia") to spread the truth (Trakhtenberg 2007:127).

The results from "Journalism in Change" confirm this picture of Russian journalism as a social field in close relation to political power. The "missionary" ambition has for those in opposition to power moved to blogs in LiveJournal and other social media platforms, and self-censorship is important for journalists in the daily work to adapt to editorial and political limits in newsrooms.

In Poland journalism carries a long tradition of resistance towards power from the long period of foreign rule during the 19th century, the world wars and 45 years of communist rule and influence from the Soviet Union (Stepinska et al., 2012:255). All the time journalists worked with a double strategy; to create some space within the official media and to establish unofficial media independent from the state. This long tradition developed strongly during the years of Solidarność in the 1980s, and still influences the position of journalism as a social field. The results are clear from the project regarding the double relation to politics – both a strong political interest among journalists and a strong sense for independence from politics.

Sweden has had a long and peaceful development for 200 years, a political history of compromise and cooperation between popular movements and political power. This has created a strong corporatism still visible in organizations within the field of journalism organizing almost all interest groups (unions, associations of employers etc.). In this organized society, journalism has developed autonomy both as a social field and for individual journalists. But at the same time it is clear in the results that journalists feel threatened by commercial pressure. This corporative system works in relation to society and politics, but not in relation to the demands from the media market.

Bourdieu concluded that the field of journalism is structured between the field of politics and economics (Bourdieu, 2005:41). But the position of journalism between these fields is very much dependent on history and culture of the national context. Journalism in Poland, Russia and Sweden gives good examples of this. But it is important to remember that there is not only continuity, but also change. Cultures are not given forever, and the changes in the media systems over the last 15–20 years also change the social position of journalism – gradually away from the pole of politics towards the pole of economics in all three countries.

Political parallelism and professional autonomy

Another way to describe the relationship between the social fields we are dealing with is the notion "political parallelism" (Hallin and Mancini, 2004:26). Earlier research shows clear differences in the intimacy of relations between the political system and journalism, in how these fields are intertwined (see Chapter 1). The results from the project confirm the importance of this basic relation and the clear differences between the countries.

Some of the results from Russia suggest that this political influence can be balanced by commercial strength. In an interview a journalist in a Russian magazine emphasized their economic independence in relation to the state. Also in the survey, there is a correlation between influence on content of the journalists and of the other parts of the media company and owners – in contrast to political influence. An interpretation of this is that when the media company gets stronger, journalists feel more freedom in relation to political power. This indicates commercial strength as one positive factor for autonomy in relation to political power.

But in Poland and Sweden, the opinions among journalists are the opposite. Commercial demands are experienced as constraints to professional autonomy for journalists, including also harder labour conditions being a threat towards quality in journalism. This gives clear negative correlations between commercial influence and professional autonomy.

10.5 The de-professionalization of journalism?

Research has described a process of professionalization of journalism (Shoemaker and Reese, 1996:93; Waisbord, 2013). In Sweden this process was strong during the 1950s and 1960s under the influence of expanding public service and the eroding system of party press (Djerf-Pierre and Weibull, 2001). Professional education with a strong emphasis on independent journalists was introduced in the 1960s (Gardeström, 2011). In Poland and Russia the same development came after 1989 – a rapid transition from a media system controlled by the communist party to a system with a more independent role for journalists and media companies (Dobek-Ostrowska, 2012a; Anikina, 2012b).

Professionalization has been described as part of the modernization of society, as a result of labour division and specialization in advanced occupations (Giddens, 2007:273). This process has been studied by sociologists to define what a profession is:

> "(…) a set of institutions which permit the members of an occupation to make a living while controlling their own work." (Freidson, 2001:17)

These institutions are for example a higher education and professional institutions controlling professionals and a kind of qualifying credentials created by the occupation to defend the standards and ethics. This gives the professionals a kind of knowledge monopoly – a special status in the labour force. To motivate this privileged position, the professions also create an ideology of altruism, to do good work is more important that financial gain (Freidson, 2001: 127–129). Classical professions are medical doctors, lawyers and priests.

Professionalization is not a question of fulfilling a list of criteria – it is about the direction of development (Tumber and Prentoulis, 2005:59). Journalism has never been regarded to be a full profession. In fact, it would be in conflict with freedom of expression to demand some kind of certificate to permit somebody to write in newspapers. Instead, journalism has been described as a profession with many of the criteria for a profession, but not as a full profession due to the lack of "closure" towards other groups (McQuail, 2010:287).

The control of work is a question of different kinds of logics. Freidson defines three kinds of logics to control specialized work in general, three ideal types that are mixed in reality. From these types it is possible to discuss the relationship between different logics (2001:179):

- The bureaucratic logic, control by rules and laws.
- The market logic – control based on market mechanisms.
- The professional logic – when professionals control their own work following standards given by the profession.

The history of professionalization of journalism involves all these kinds of logics, and it is a history of increasing professional logic. But the direction of the process is not given forever. In post-industrial society researchers also find tendencies of de-professionalization, when other kinds of logics become stronger and the professional logic gets weaker. It is also a question about the strength of structures and institutions protecting the autonomy of journalism, making it possible to defend a professional logic (Witschge and Nygren, 2009; Waisbord, 2013).

In the project "Journalism in Change," we can see clear signs of a de-professionalization of journalism in all three countries. This development has different points of departure, in Sweden journalism had a very strong position as a profession, and Poland and Russia experienced a period of rapid professionalization and growing autonomy in the 1990s. But the direction in development is now similar in all three countries:

- Knowledge monopoly for journalists is gradually weaker as other kinds of communication in the Internet are being developed. The professional control of journalists is challenged by open participation.
- Professional institutions for journalists are becoming weaker in Poland and Russia, but also to a certain degree among young journalists in Sweden.
- The borders between journalism and other kinds of media work and media content are softer – for example in relation to PR and "content marketing."
- Common standards and ethics have a weak support, at least in Russia and Poland. Journalism is becoming more heterogeneous.
- Market logic is gaining influence, professional logic and the public service-orientation is gradually losing ground.

Professionalization is still important, especially in Sweden, where journalists have a strong position in media companies and in society. Also in Poland and Russia the ideology of professionalism is used by journalists to achieve a larger degree of autonomy and to defend freedom of the press. The survey and interviews give many examples of this strive for professional autonomy – in values of journalists and in experiences from daily work.

Tendencies towards de-professionalization are not mainly about decreasing quality in journalism, at least not in the first step. It is about the strength of the profession to defend common standards and ethics, to achieve a degree of autonomy making it possible to follow the professional logic when it comes to conflicts with other logics.

Globalization of the professional logic in journalism makes journalists think more similar when it comes to news, reporting practices and standards in the profession (Waisbord 2013: 227–229). Ideals may differ, but professional logics are used by journalists to gain control in their field of work, to achieve more autonomy for the profession. News values and ethical norms are parts of this professional logic in journalism, and help journalists to gain control and autonomy. But the way this is done is shifting in hybrid forms of professional cultures.

Professionalism remains a key part of journalism in a societal perspective. The professional logic emphasizes the public service duties of journalism, the audience's and society's right to know. Other kinds of logics emphasize other needs like commercial success or the needs of political elites. The "Journalism in Change" project shows tendencies towards a de-professionalization in Poland, Russia and Sweden – but also how journalists in all three countries defend professional logic and professional values. As formulated by Waisbord in "Reinventing Professionalism":

> "Professionalism offers a refuge to protect news from being vulnerable to governments and markets uninterested in the pursuit of critical rationality, accountability and transparency. Professionalism is necessary for journalism to be a counterweight to power. Professionalism is needed to nurture an occupational culture committed to democratic goals: to scrutinize abuses, to give voice to citizens (particularly those systematically ignored by power), to monitor the actions of governments and markets and to promote social empathy and solidarity." (Waisbord, 2013:226)

Hypotheses and the future

At the beginning of our research we posed four hypotheses. We predicted many similarities in the journalistic cultures as a consequence of market influence and liberal ideals (H1). But we also believed that nationally rooted traditions from history and culture still influence journalistic cultures deeply and preserve differences (H2). We thought that media development makes the profession weaker at the collective level (H3). However, at the same time, new technology can strengthen the position of the individual journalist (H4).

In general, the results of the project confirm these hypotheses. The journalistic cultures are in some aspects moving closer to each other. But still there are deeply rooted national differences between journalism in the three countries. Also, even if ideals are getting more similar, the conditions of journalistic practice and pressure from other social fields create clear differences in how ideals and values are fulfilled.

Technology and commercialization are important factors changing journalism and the media systems in all three countries. The professional culture is under pressure, and in many ways transforming to be more heterogeneous. At the same time individual journalists are offered new tools and new platforms for publishing, giving individual journalists a stronger position.

What does this say about the future? The results of the "Journalism in Change" project lead us to suppose that the profession would change deeply in all three countries. In the case of Sweden and Poland the economic environment of the media will produce a stronger commercialization, but also in Russia this process will be more visible. The elder professional generation of journalists in Sweden and Poland will be retired and a younger generation will replace it with slightly changing ideals and practices. Today, the young generation of audience use traditional media in a very limited way. It means that a crisis of these media will deepen, which is why journalism has to search for new ways of reaching audiences. For Russia, the development is also closely connected to the political situation; there is not only a commercial challenge but also political power imposing narrow limits upon journalism.

We shall close this report by giving a voice to some of the journalists in the three countries expressing both fear and confidence, looking forward into the future of journalism:

> "(…) There *will be no printed newspapers* [in the future – BDO and GN] but I am not convinced because printed papers still make the biggest profits from advertisements and it is the main source of financing papers." [PR10, female, over 50 years, regional newspaper, journalist]

> "(…) broadcasting television *will gradually lose its popularity.* Television *will move on the web in full.* Of course, as always, the demand will be for the materials and texts of good quality, and in 20 years things will stay the same. This is why, despite the huge number of amateur blogs, photos, and other "popular" content, there will always be demand for high-quality journalism. "[RR 2, female, 24 years old, state TV, editor]

> "Newspapers *will lose much in favour of the combined media* tablets, computers and phones. There will be many more moving pictures, but also some writing. I think there will be much shorter texts: shortened, but it will be supplemented with a number of people who work with longer and more in-depth assignments and reports. I don't think the

interest in stories will disappear in these years. But it will be a shrinking market for the pure written newspaper story." [SR 11, male, 54 years old, freelancer and photographer]

"The Internet is a democratizing force in the world and we are actually only at the beginning of that change. *We will see more new phenomena and new technology that will revolutionize many areas* – this also applies to journalism." [SR 20, male, 40 years old, web editor radio].

References

About the Swedish Union of Journalists. Retrieved April 23, 2014 from https://www.sjf.se/about.

Allan, S. (ed.) (2010). *The Routledge Companion to News and Journalism*. London and New York: Routledge.

Allan, S. (2009). Histories of Citizen Journalism. In: S. Allan and E. Thorsen (eds.). *Citizen Journalism: Global Perspectives (Global Crises and the Media)*. New York: Peter Lang Publishing, 17–32.

Allan, S. and E. Thorsen (eds.) (2009). *Citizen Journalism: Global Perspectives*. New York: Peter Lang.

Almanach Mediów 2012 (2013). Warszawa: Press.

Altmeppen, K. -D. (2008). The Structure of News Production: the Organizational Approach to Journalism Research. In: D. Löffenholz and D. Weaver (eds.). *Global Journalism Research: Theories, Findings, Future*. Oxford: Blackwell Publishing.

Anderson, C.W., E. Bell and C. Shirky (2013). *Post-Industrial Journalism: Adapting to the Present*. New York: Colombia School of Journalism.

Anderson, C. and M. Schudson (2009). Objectivity, Professionalism and Truth Seeking in Journalism. In: K. Wahl-Lorgensen and T. Hanitzsch (eds.). *The Handbook of Journalism Studies*. New York: Routledge, 88–101.

Andersson, U. (2009). *Journalister och deras publik: förhållningssätt bland svenska journalister* [Journalists and their audience: how journalists relate to them]. Göteborg: Institutionen för journalistik och masskommunikation (JMG), Göteborgs universitet.

Anikina, M. (2012a). Journalist and Journalist's Culture in Russia. In: E. Vartanova (ed.). *World of Media 2012 Yearbook of Russian Media and journalism Studies*, Moscow, Moscow State University: Media Mir, 243–268.

Anikina, M. (2012b). Journalists in Russia. In: G. Nygren (ed.). *Journalism in Russia, Poland and Sweden – traditions, cultures and research*. Stockholm: Södertörns University, 13–31.

Anikina, M. (2012c). The Smolensk tragedy in Russian media: sociological analysis. In A. Gralczyk, M. Laskowska, P. Drzewiecki (eds.). *Media wobec śmierci, tom 2*. Warszawa, 235–247.

Anikina, M, B. Dobek-Ostrowska G. Nygren (eds.) (2013). *Journalists in Three Media Systems. Polish, Russian and Swedish journalists about values and ideals, daily practice and the future.* Moscow: Moscow State University.

Anikina, M., B. Dobek-Ostrowska and G. Nygren (2013). A cross- national survey. In A. Anikina, B. Dobek-Ostrowska and G. Nygren (eds.). *Journalists in Three Media Systems. Polish, Russian and Swedish journalists about values and ideals, daily practice and the future.* Moscow: Moscow State University, 30–34.

Anikina, M. and E. Johansson (2013). Russian journalists: inclination towards moderate evaluations. In M. Anikina, B. Dobek-Ostrowska and G. Nygren (eds.). *Journalists in Three Media Systems. Polish, Russian and Swedish journalists about values and ideals, daily practice and the future.* Moscow: Moscow State University, 69–114.

Appelgren, E. and G. Nygren (2014). Data journalism in Sweden. Introducing New Methods and Genres of Journalism into "Old" Organizations. *Digital Journalism* Vol. 2 (3), 394–405.

Appelberg, J., E. Johansson, G. Nygren and P. Baranowski (2014). Social Media in Professional Work of Polish, Russian and Swedish Journalists. *Journal of Print and Media Technology Research.* IARIGAI, Darmstadt, Vol. 3 (2), 107–118.

Asp, K. (ed.) (2012) *Svenska journalister: 1989–2011* [Swedish journalists 1989–2011]. Göteborg: Institutionen för journalistik, medier och kommunikation (JMG), Göteborgs universitet.

Asp, K. (ed.) (2007). *Den svenska journalistkåren* [Swedish journalist]. Gothenburg: JMG Gothenburg University.

Asp, K. and B. Johansson (2007). Makten over journalistiken [Power over journalism]. In:, K. Asp (ed.) *Den svenska journalistkåren* [The Swedish Journalists]. Göteborg: JMG/Gothenburg University, 163–172.

Bajomi – Lázár, P. (2014). *Party Colonisation of the Media in Central and Eastern Europe.* Budapest: CEU Press.

Bajomi – Lázár, P. (2014). How the internet changed/s journalism: some trends in the 'West' and 'East'. In: M. Głowacki, E. Lauk, A.Balčytienė (eds.) (2014). *Journalism that Matters. Views from Central and Eastern Europe.* Frankfurt am Main: Peter Lang Edition, 61–78.

Bak, J. S. (2011). Introduction. In: J.S. Bak and B. Reynolds (eds.). *Literary Journalism Across the Globe: Journalistic Traditions and Transnational Influences.* University of Massachusetts Press, 1–20.

Balčytienė, A., E. Lauk and M. Głowacki (2014). Roller coasters of social change, democracy and journalism in Central and Eastern Europe. In: M. Głowacki, E. Lauk and A.Balčytienė (eds.). *Journalism that Matters. Views from Central and Eastern Europe*. Frankfurt am Main: Peter Lang Edition, 9–20.

Baldin, E., and D. Borodin (2012). Wave 6: Issledovanie auditorii servisov Web 2.0 v Rossii. [Wave 6: the research of Web 2.0 services in Russia] In: A. Shashkin, I. Devyatko and S. Davydov (eds.) *Onlajn issledovanija v Rossii. 3.0.*[Online research in Russia. 3.0.]. Moscow: Publishing house Codex, 197–208.

Benson, R. and E. Neveu (2005). *Bourdieu and the Journalistic Field*. London: Polity.

Bergström, A. (2011). Social media in campaigning – citizens and politicians in the 2010 Swedish election. *Central European Journal of Communication*, vol. 4, 2 (7), 241–258.

Berkowitz, D. A. (2009). Reporters and Their Sources. In: T. Hanitzsch and K. Wahl-Jorgensen (eds.) *The Handbook of Journalism Studies*. New York and London: Routledge, 102–115.

Björklund, F. and J. Rodin (2009). *Det nya Östeuropa – stat och nation i förändring* [The new Eastern Europe – state and nation in change]. Lund: Studentlitteratur.

Blumler, J. and M. Gurevitch (1995). *The Crisis of Public Communication*. London: Routledge.

Bodrunova, S., and A. Litvinenko (2012). Hybridization of the Media System in Russia: Technological and Political Aspect. In: E. Vartnova (ed.) *World of Media. Yearbook of Russian Media and Journalism Studies*. Moscow: Faculty or Journalism Lomonosov State University, 37–49.

Bohere, G. (1984). *Profession: Journalist. A study of the working conditions of journalists*. Geneva: International Labor Office.

Bourdieu, P. (2005) The Political Field, the Social Science Field and the Journalistic Field. In: R. Benson, R. and E. Neveu (eds.) *Bourdieu and the Journalistic Field*. London: Polity, 29–27.

Broersma, M., and Ch. Peters (2013). Introduction – Rethinking Journalism: The Structural Transformation of a Public Good. In: C. Peters and M. Broersma (eds.) *Rethinking Journalism: Trust and Participation in a Transformed News Landscape*. London, New York: Routledge.

Burke, E. (1989). *On Symbolsand Society*. Chicago: University of Chicago Press.

Cafaggi, F., F. Casarosa and T. Prosser (2012). *The regulatory quest for free and independent media. Mediadem Report*. Retrieved April 22, 2014 from http:// www. mediadem.eliamep.gr/wp-content/uploads/2012/09/D3.2.pdf.

Castells, M. (1996) *The Information Age: Volume 1: The Rise of the Network Society.* Oxford: Blackwell Publishers.

Chadwick, A. (2013). *The Hybrid Media System. Politics and Power*. Oxford: Oxford University Press.

Chalaby, J. (1996) Journalism as an Anglo-American Invention. *European Journal of Communication.* 11(3), 303–326.

Champagne, P. (2005). The Double Dependency: The Journalistic Field between Politics and Markets. In: R. Benson and E. Neveu (eds.). *Bourdieu and the Journalistic Field*. London: Polity, 48–63.

Chappel, B. (2012). *Social Network Analysis Report – Demographic – Geographic and Search Data Revealed. IGNITE Social Media*. Retrieved May 7, 2014 from http://www.ignitesocialmedia.com/social-media-stats/2012-social-network-analysis-report/#Livejournal.

Christians, G. C. et al. (eds.) (2009). *Normative theories of the media: journalism in democratic societies.* Chicago: University of Illinois Press.

Clifford, G. C. et al. (eds.) (2009). *Normative theories of the media: journalism in democratic societies.* Chicago: University of Illinois Press.

Craft, S. and C. N. Davis (2013). *Principles of American Journalism: An Introduction*. London: Routledge.

Craft, S., Davis C. N. (2013). *Principles of American Journalism: An Introduction*. London: Routledge.

Croteau, D, and W. Hoynes (2000). *Media/Society. Industries, Image, and Audience*. Thousand Oaks. London and New Delhi: Pine Forge Press.

Curry, J. (2013). Profession in transition: Journalistic professionalism and its changes in Central and Eastern Europe. Interview. *Central European Journal of Communication*. Vol. 6, no 1 (10), 144–150.

Curry, J. (2008). Poland: The Politics of "God's" Playground. In: S.L. Wolchik, J. Curry (eds.), *Central and Eastern European Politics from Communism to Democracy*. New York: Rowman and Littlefield Publishers, Inc., 165–190.

Curry, J. (1990). *Poland's Journalists. Professionalism and Politics*. Cambridge: Cambridge University Press.

Curry, J. (1980). Polish Journalists in the Policy-Making Process. In: R. Kenet (ed.), *Background to Crisis: Policy and Politics in Gierek's Poland*, Boulder: Westview Press, 177–198.

Daily media reach in Sweden 2012 (2012). Retrieved May 7, 2014 from http://www.nordicom.gu.se/sites/default/files/mediefakta-dokument/Mediebarometern/english_bar_2012_0.pdf.

Davies, N. (2008). *Flat Earth News*. London: Chatto and Windus.

Deaths of Journalists in Russia (2014). Retrieved May 7, 2014 from http://www.Journalist-in-russia.org/

Democracy Index 2013 (2013). Retrieved April 23, 2014 from http://www.eiu.com/Handlers/WhitepaperHandler.ashx?fi=DemocracyIndex_2013WEB-2.pdf&mode=wp&campaignid=Democracy0814.

Democracy index 2012 (2012). Democracy at a standstill. A report from The Economist Intelligence Unit. Retrieved May 7, 2014 from https://portoncv.gov.cv/dhub/porton.por_global.open_file?p_doc_id=1034.

Democracy Ranking of the Quality of Democracy 2013 (2013). Retrieved May 7, 2014 from http://democracyranking.org/wordpress/?page_id=738.

de Smaele, H. (2010). In Search of a Label for the Russian Media System. In: B. Dobek-Ostrowska, M. Głowacki, K. Jakubowicz, M. Sükösd (eds.). *Comparative Media Systems. European and Global Perspective*. Budapest – New York: CEU Press, 41–62.

Deuze, M. (2007). *Media Work. Digital Media and Society Series*. Cambridge: Polity Press.

Dinan, W. and D. Miller (2009). Journalism, Public Relation and Spin. In: T. Hanitzsch and K. Wahl-Jorgensen (eds.). *The Handbook of Journalism Studies*. New York and London; Routledge, 250–264.

Djerf-Pierre, M. and L. Weibull (2001). *Spegla, granska, tolka. Aktualitetsjournalistik i svensk radio och TV under 1900-talet* [Mirror, scrutinize and interpret. News journalism in Swedish radio and TV during the 20th century]. Stockholm: Prisma.

Dobek-Ostrowska, B. (2012a). Italianization of the Polish Media System? Reality and Perspective. In: D. Hallin, and P.Mancini (eds.). *Comparing media systems beyond the western world*. New York: Cambridge University Press, 26–50.

Dobek-Ostrowska, B. (2012b). Journalists in Poland – some previous research. In: G. Nygren, (ed.). *Journalism in Russia, Poland and Sweden – traditions, cultures and research*. Stockholm: Södertörns University, 32–51.

Dobek-Ostrowska, B. (2011a). *Polski system medialny na rozdrożu. Media w polityce, polityka w mediach* [Polish Media System at a Crossroads. Media in Politics, Politics in Media]. Wrocław: Wrocław University Press.

Dobek-Ostrowska, B. (2011b). Political parallelism or political bias? Consequences for the quality of democracy in Poland. In: B. Dobek-Ostrowska and M. Głowacki (eds.). *Making Democracy in 20 Years. Media and Politics in Central and Eastern Europe*. Wrocław: University of Wrocław Press, 193–211.

Dobek-Ostrowska, B. (2004). *Media masowe i aktorzy polityczni w świetle studiów nad komunikowaniem politycznym* [Mass media and political actors in contex of political communication studies]. Wrocław: University of Wrocław Press.

Dobek-Ostrowska, B., M. Głowacki and M. Kuś (2014). Context Factor for Media Self-Regulation and Accountability. In: Fengler, S., G. Mazzoleni, S. Russ-Mohl, C. Porlezza and T. Eberwein (eds.). *Journalists and Media Accountability: An International Study of News People in the Digital Age*. New York: Peter Lang, 149–166.

Dobek- Ostrowska, B., P. Barczyszyn, A. Michel and P. Baranowski (2013). Polish journalists – from politics to market. In: M. Anikina, B. Dobek-Ostrowska and G. Nygren (eds.). Journalists in Three Media Systems. Moscow: Faculty of Journalism – University of Lomonosov, 35–68.

Dobek-Ostrowska, B., P. Barczyszyn and A. Michel (2013). Zmiana w dziennikarstwie. Kultura zawodowa polskich dziennikarzy [Journalism in Change: Journalism Culture of Polish Journalists]. *Studia Medioznawcze*, Nr. 1 (52), 11–27.

Donsbach, W. (2010). Journalists and their professional identities. In: S. Allan (ed.). *The Routledge Companion to News and Journalism*. London and New York: Routledge, 33–48.

Eberwain, T., S. Fengler, E. Lauk, and T. Leppik-Nork (eds.) (2011). *Mapping Media Accountability – in Europe and Beyond*. Koln: Herbert von Halem Verlag.

Erzikova, E. and W. Lowrey (2011). Managed mediocrity Experienced journalists' perspectives on the "pampers generation" of Russian reporters. *Journalism Practice*, 6:2: 264–279.

Engblom, L. Å. (2001). Varför professionalisering? [Why professionalization?]. In A. Lindblom Hultén (ed.) *Journalisternas bok 1901–2001* (The Journalist's book 1901–2001) Stockholm: Svenska journalistförbundet.

Esaiasson, P., M. Gilljam, H. Oscarsson, and L. Wägnerud (2003). *Metodpraktikan – konsten att studera samhälle, individ, marknad* [Practices in methods – the art of studying society, individuals and market]. Stockholm: Norstedts Juridik.

Etling, B., K. Alexanyan, J. Kelly, R. Faris, J. Palfrey and Gasser, U. (2010). *Public Discourse in the Russian Blogosphere: Mapping RuNet Politics and Mobilization*. Berkman Center Research Publication, No. 2010–11, October 19, 2010. Retrieved: May 7, 2014 from http://cyber.law.harvard.edu/teaching/ilaw/2011/sites/teaching/images/Public_Discourse_in_the_Russian_Blogosphere_2010.pdf.

Evett, J. (2003). The Sociological Analysis of Professionalism. *International Sociology*, 18(2), 395–415.

Evett, J. and M. Aldridge. (2003). Rethinking the concept of professionalism: the case of journalism. *British Journal of Sociology, 54*(4), 547–564. Retrieved March 20, 2014 from https://www.um.es/ESA/papers/Rn15_23.pdf.

Fengler, S., G. Mazzoleni, S. Russ-Mohl, C. Porlezza, and T. Eberwein (eds) (2014), *Journalists and Media Accountability: An International Study of News People in the Digital Age*. New York: Peter Lang.

Fenton, N. (ed.) (2010). *New Media, Old News: Journalism and Democracy in the digital age*. London: Sage.

Fernández-Quijada (2014). A Golden Decade. Exploring Internationalization in Nordic Communication Research. *Nordicom Review*, vol. 34, 1 (14), 135–150.

56 Journalists Killed in Russia since 1992/Motive Confirmed (2014). Retrieved April 20, 2014 from http://cpj.org/killed/europe/russia/.

Florida, R. (2002). The Rise of the Creative Class. And How It's Transforming Work, Leisure and Everyday Life. New York: Basic Books.

Fomicheva, I. (2012). *Sotsiologiya SMI* [Media sociology]. Moscow: Aspekt Press.

Franks, S. (2005). Lacking a clear narrative: Foreign News Reporting after the Cold War, *The Political Quarterly*, 76 (1), 91–101.

Freidson, E. (2001). *Professionalism: The third logic.* Cambridge: Polity Press.

Furhoff, L. (1974). *Makten över medierna* [The power over media].Stockholm: Bo Cavefors bokförlag.

Gade, P. (2008). Journalism guardians in a time of great change: News editors' perceived influence in integrated news organizations. *Journalism and Mass Communication Quarterly*, 85, 371–392.

Galtung, J. (1999). State, Capital, and the Civil Society: The Problem of Communication. In: R. Vincent, K. Nordenstreng and M. Traber (eds.). *Towards Equity in Global Communication: MacBride Update*. Cresskill, NJ: Hampton Press, 3–21.

García-Avilés, J., A. Kaltenbrunner and K.Meier (2013). *Media Convergence Revisited: Lessons Learned on Newsroom Integration in Austria, Germany and Spain*. Paper presented at the International conference "Future of Journalism", Cardiff, UK, 2013.

Garcia de Torres, E., L. Yezerska, A. Rost, M. Calderin, M. Rojano, C. Edo, E. Sahid, P. Jerónimo, C. Arcila,, A. Serrano J. Badillo and L. C. Alfonso (2011). *See You on Facebook or Twitter? The Use of Social Media by 27 News Outlets from 9 Regions in Argentina, Colombia, Mexico, Peru, Portugal, Spain and Venezuela.* Paper presented at 12th International Symposium for Online Journalism, Austin, 2011.

Gardeström, E. (2011) *Att fostra journalister: journalistutbildningens formering I Sverige 1944–1970 (To foster journalists: formation of journalism education in Sweden 1944–1970)* Göteborg: Daidalos.

Giddens, A. (2007) *Sociologi, fjärde omarbetade upplagan* [Sociology, fourth revised edition] Lund: Studentlitteratur.

Ghersetti, M., and L. Weibull (2009). The Swedish Journalism Education Landscape. In: G. Terzis (ed.). *European Journalism Education*. Bristol and Chicago: Intellect.

Global Corruption Barometer 2010/2011. Retrieved: February 22, 2014 from http://www. transparency.org/gcb201011/results.

Głowacki, M. (2013). Dziennikarze polscy w świetle studiów porównawczych systemów odpowiedzialności mediów w Europie [Polish journalists in comparative analysis of media accountability systems in Europe]. *Studia Medioznawcze*, 1 (52), 29–44.

Głowacki, M. E. Lauk, A.Balčytienė (eds.) (2014). *Journalism that Matters. Views from Central and Eastern Europe*. Frankfurt am Main: Peter Lang Edition.

Gorny, E. (2011). *Understanding the Real Impact of Russian Blogs. Analysis. Russian analytical digest 69/09*. Center for Security Studies CSS Swiss Federal Institute of Technology ETH, Zurich. Retrieved: May 7, 2014 from www.css. ethz.ch/publications/pdfs/RAD-69–8-11.pdf.

Gorny, E. (2007). *The Russian Internet: Between Kitchen-Table Talks and the Public Sphere. ARTMargins, online*. Retrieved: May 7, 2014 from php/2-articles/145-the-russian-internet-between-kitchen-tabletalks-and-the-public-sphere.

Gorny, E. (2004). *Russian LiveJournal: National Specifics in the Development of a Virtual Community.* Version 1.0 of 13 May 2004. Retrieved February 2, 2014 from http://www.ruhr-uni-bochum.de/russ-cyb/library/texts/en/journal/journal_1.htm.

Gritsanov A. et al. (2003). *Sotsiologiya: Entsiklopediya* [Sociology: Encyclopedia]. Minsk: Interpresservis, Knizhnyy Dom.

Gros, P. and K. Jakubowicz (2013). *Media Transformations in the Post-Communist World: Eastern Europe's Tortured Path to Change.* Lexington Books.

Grushin, B.A. and L.A. Onikov (1980). *Massovaya informatsiya v sovetskom promyshlennom gorode* [Mass information in a Soviet Industrial town]. Moscow: Izdatel'stvo politicheskoy literatury.

Gudkov L. (2014). *Massovye otsenki i nastroeniya v 2013 godu* [Mass appraises and mood in 2013]. Retrieved April 23, 2014 from http://www.levada.ru/sites/default/files/gudkov_l._dinam ika_nastroeniy_2013.pdf.

Gustafsson, K-E., and P. Rydén (2010) *A history of the press in Sweden.* Gothenburg: Nordicom.

Hadamik, K. (2005). Between East and West or Simply "Made in Poland". The Many Different Styles of Today's Polish Journalism. In: S. Hoyer, H. Pottker (eds.). Diffusion of News Paradigm *1850–2000*, Goteborg: Nordicom, 211–224.

Hallin, D. and P. Mancini, P. (eds.) (2012). *Comparing Media Systems Beyond the Western World.* New York:Cambridge University Press.

Hallin, D. and P. Mancini (2011). Comparing Media Systems in Central and Eastern Europe. Interview with D. C. Hallin and P. Mancini on models of media systems, journalism, new technologies and an Eastern European Perspective in comparative media systems. *Central European Journal of Communication,* Vol. 4, no 2 (7), 317–322.

Hallin, D. and P. Mancini (2004). *Comparing Media Systems. Three models of Media and Politics.* Cambridge: Cambridge University Press.

Hanitzsch, T. (2009). Comparative Journalism Studies.in T. Hanitzsch and K. Wahl-Jorgensen (eds.). The Handbook of Journalism Studies. New York and London: Routledge, 413–427.

Hanitzsch, T. (2007). Deconstructing Journalism Culture: Toward a Universal Theory, *Communication Theory.* (17), 367–385.

Hanitzsch, T., J. Seethaler, E. Skewes, M. Anikina, M. Berganza, R. Cangoz, I. Coman, M. Hamada, B. Hanusch, F. Karadjov, C. Mellado, C. Moreira, S.V. Mwesige, P. Plaisance, P.L. Reich, Z. Noor and D.V. Yuen, K.V. (2012). Worlds of journalism: Journalistic cultures, professional autonomy and perceived influence across 18 nations. In D. Weaver and L. Willnat (eds.) (2012). *The Global Journalist in the 21st Century.* New York and London: Routledge, 473–495.

Hanitzsch, T., Hanusch, F., Mellado, C., Anikina, M., Berganza, R., Cangoz, I., Coman, M., Hamada, B., Hernandez, M.-E., Karadjov, C.D., Moreira, S.-V., Mwesige, P.G., Plaisance, P. L., Reich, Z., Seethaler, J., Skewes, E. A., Noor, D. V., Kee Wang Yuen, E. (2011). Mapping Journalism Cultures across Nations: a Comparative Study of 18 Countries. *Journalism Studies*. Routledge, Vol. 12 (3), 273–293.

Hanitzsch, T. et al. (2010). Modeling Perceived Influences on Journalism: Evidence from a Cross-national Survey of Journalists. *Journalism and Mass Communication Quarterly.* 87 (1), 5–22. Retrieved from http://eprints.qut.edu.au/68292/1/2010_-_Hanitzsch_et_al_-_JMCQ_-_Modeling_Perceived_Influences.pdf.

Harro-Loit, H., E. Lauk, H. Kuutti, U. Loit (2012). Professional autonomy in journalism as a factor for safeguarding freedom of expression: A comparative perspective. In: *Media freedom and independence in 14 European countries: A comparative perspective.* European Commission; Seventh framework programme. Retrieved April 14, 2014 from www.mediadem.eliamep.gr.

Harvey, K. (ed.) (2014). Social Media, Definition and Classes. In: K. Harvey and J.G. Golson, (eds.) *Encyclopedia of Social Media and Politics.* Thousand Oaks: Sage.

Hedman, U., and M. Djerf-Pierre (2013). The Social Journalist: Embracing the Social Media Life or Creating a New Digital Divide? *Digital Journalism.* Vol. 1(3), 368–385.

Heinonen, A. (2011). The Journalist's Relation with Users: New Dimensions to Conventional Roles. In J. B. Singer, A. Hermida, D. Domingo, A. Heinonen, S. Paulussen, T. Quandt, Z. Reich, and M. Vujnovic (eds.). *Participatory Journalism: Guarding Open Gates at Online Newspapers.* Oxford: Wiley-Blackwell, 34–55.

Hermida, A. (2012). Social Journalism: Exploring How Social Media is Shaping Journalism. In: E. Siapera and A. Veglis (eds.). *The Handbook of Global Online Journalism.* New York: John Wiley Sons, Inc, 309–328.

Hermida, A. (2011). Fluid Spaces, Fluid Journalism. In: J. B. Singer, A. Hermida, D. Domingo, A. Heinonen, S. Paulussen, T. Quandt, Z. Reich, and M. Vujnovic (eds.). *Participatory Journalism: Guarding Open Gates at Online Newspapers.* Oxford: Wiley-Blackwell, 177–191.

Hiebert, R., and P. Gross (2003). Remedial Education: The Remaking of Eastern European Journalists. In: R. Frohlich and Ch. Holtz-Bacha (eds.). *Journalism Education in Europe and North America. An International Comparison.* Cresskill, NJ: Hampton Press, 257–283.

Hök. J. (2012). Swedish journalism- a struggle for autonomy. In: G. Nygren, (ed.). *Journalism in Russia, Poland and Sweden – traditions, cultures and research.* Stockholm; Södertörns University, 53–72.

Humphreys, P. (1996). *Mass Media and Media Policy in Western Europe*. Manchester: Manchester University Press.

Internet Retrieved May 7, 2014 from http://www.tns-global.ru/services/media/media-audience/internet/information/?arrFilter_pf[YEAR]=2012&set_filter=%D0%9F%D0%BE%D0%BA%D0%B0%D0%B7%D0%B0%D1%82%D1%8C&set_filter=Y.

Internet and Facebook Usage in Europe. Internet and Facebook Stats. Usage and Population Statistics (2013). Retrieved: May 7, 2014 from http://www. internetworldstats.com/stats4.htm#europe.

Jakubowicz, K. (2009). *A new notion of media*. Retrived October 10, 2–14 from http://www.coe.int/t/dghl/standardsetting/media/doc/New_Notion_Media_en.pdf.

Jakubowicz, K. (2008). Riviera on the Baltic? Public Service Broadcasting in Post-communist Countries. In: B. Dobek-Ostrowska and M. Glowacki (eds.). *Comparing Media Systems in Central Europe. Between Commercialization and Politicization*. Wroclaw; University of Wroclaw Press, 41–54.

Jakubowicz, K. (2002). Media in Transition: The Case of Poland. In: M. E. Price, B. Rozumilowicz and S. G. Verhulst, (eds.) *Media Reform. Democratizing the Media, Democratizing the State.* Routledge, London; New York, 203–231.

Jenkins, H. (2006). *Convergence Culture: Where Old and New Media Collide.* New York: New York University Press.

Jenkins, H., Ford, S. and J. Green (2013). *Spreadable Media: Creating Value and Meaning in a Networked Culture.* New York: New York University Press.

Jenkins, H., R. Purushotma, M. Weigel, K. Clinton, and A. J. Robison (2009). *Confronting the Challenges of Participatory Culture Media Education for the 21st Century*. Cambridge, Massachusetts, London: The MIT Press.

Johansson, E. (2014). The Blog Platform LiveJournal as a Professional Tool of Russian Journalists. *Baltic Worlds*. CBEES, Södertörn University, Stockholm, Vol. 2, 27–36.

Johansson, E. (2013). *LiveJournal Blog Platform as a Tool for Professional Work of Russian Journalists.* Paper presented at the International conference "Future of Journalism", Cardiff.

Johansson, E. and G. Nygren (2014). Russian Journalists and Social Media: Updated Traditions and New Challenges. *Central European Journal of Communication.* 7 (2), 101–118.

Kachkaeva, A. (2013). Zarjadka ot seti [Charging from a network]. In: Ch. Moeller and M. Stone (eds.). *Putevoditel po socialnym media* [Social media guide]. Predstavitel OBSE po voprsam svobody SMI [Office of the OSCE Representative on Freedom of the Media]: Vienna, 85–92.

Kak izmenilis lyudi i vlast za *20 let* [How people and authorities had been changed in 20 years] (2013) URL. Retrieved April 23, 2014 from http://www.levada.ru/ http://www.levada.ru/20-08-2013/kak-izmenilis-lyudi-i-vlast-za-20-let.

Kaplan, A. M. and M. Haenlein (2010). Users of the World, Unite! The Challenges and Opportunities of Social Media. *Business Horizons*, 53 (1), 59–68.

Karlsson, M. (2010). *Nätnyheter. Från sluten product till öppen process* [Online news. From a closed product to an open process]. Stockholm: SIMO förlag.

Kellner, P. (2010). *Number cruncher: a matter of trust.* Retrieved April 20, 2014 from http://www.prospectmagazine.co.uk/magazine/peter-kellner-yougov-trust-journalists/#.2 V MrPtvWDp.

Kiriya, I. (2014). Social Media as a Tool of Political Isolation in the Russian Public Sphere. *Journal of Print and Media Technology Research.* Darmstadt, Vol. 3 (2), 131–138.

Kiriya I. (2013). *Bernar M'ezh: Uchenyj-mnogostanochnik.Pokolenija VSE. Ucheniki ob uchiteljah* [A scholar – a multy-mashine-tool servant. Generations of HSE (Higher School of Economics). Pupils about teachers] M.: Izdatel'skij dom NIU VSE, 52–56.

Kiriya, I. (2012). The Culture of Subversion and Russian Media Landscape. *International Journal of Communication.* No. 6, 446–466.

Kiriya, I. and E. Degtereva (2010). Russian TV Market: Between State Supervision, Commercial Logic and Simulacrum of Public Service. *Central European Journal of Communication.* Vol. 3. No. 1(4), 37–51.

Kiriya, I., and A.A. Novikova (2013). Depressivnoe mediapotreblenie (issledovanie televizionnyh predpochtenij sel'skih zhitelej) [Depressive media consumption. Research of TV preferences of rural inhabitants]. *Vestnik Moskovskogo universiteta. Serija 10: Zhurnalistika* [Vestnik MSU, Series 10 Journalism], 5, 53–63.

Knight, M. and C. Cook (2013). *Social Media for Journalists: Principles and Practice.* London: Sage.

Konow Lund, M. (2013). *Moving-Image News production in Two Newsrooms.* Oslo: University of Oslo.

Kovach, B. and Rosenstiel, T. (2001). *The Elements of Journalism.* New York: Crown publishers.

Ladd, J. (1973). *Ethical Relativism.* Belmont, CA: Wadsworth.

Lasorsa, D. L., Lewis S. C. and A. E. Holton. (2012). Normalizing Twitter: Journalism Practice in an Emerging Communication Space. *Journalism Practice.* Routledge, Vol. 13 (1), 19–36.

Lauk, E. (2009). Reflections on Changing Patterns of Journalism in the New EU Countries. *Journalism Studies*, 10 (1), 69–84.

Lavrakas. P. (2013). *Encyclopedia of Survey Research Methods* Pub. date: | DOI: 10.4135/9781412963947, downloaded 2013/07/21.

Law of Russian Federation on Mass Media 27.12.1991 N 2124–1, Retrieved August 5, 2014 from http://www.democracy.ru/english/library/laws/eng_1991–1/.

Leetaru, K. H., Sh. Wang, G. Cao, A. Padmanabhan, and E. Shook (2013). Mapping the Global Twitter Heartbeat: The Geography of Twitter. *First Monday*, Vol. 18, 5–6. Retrieved May 7, 2014 from http://firstmonday.org/article/view/4366/3654.

Lee-Wright, P. (2012). The return of Hephaestus. Journalists work recrafted. In: P. Lee-Wright, A. Philips, and T. Witschge (eds.). *Changing Journalism.* London and New York: Routledge, 21–40.

Lee-Wright, P., A. Philips, and T. Witschge (eds.) (2012). *Changing Journalism.* London and New York: Routledge.

Levada Center (2013) URL: Retrieved February 27, 2014 from http://www.levada.ru.

Levy, D. and Nielsen, R. K. (2010). *The Changing Business of Journalism and its Implications for Democracy*. Oxford: Reuters Institute for the Study of Journalism.

Lewis, S. C. (2012). The Tension between Professional Control and Open Participation. *Information, Communication and Society.* Vol. 15 (6), 836–866.

Lor, P. (2011). *International and Comparative Liberianship*. Chapter 4: Methodology in comparative studies, 24. Retrieved September 23, 2014 from http://pjlor.files.wordpress.com /2010/06/chapter-4-draft-2011-04-20.pdf.

Lunden, I. (2014). *Instagram Is The Fastest-Growing Social Site Globally, Mobile Devices Rule Over PCs For* Access. TechCrunch. Retrieved June 6, 2014 from http://techcrunch.com/2014/01/21/instagram-is-the-fastest-growing-social-site-globally-mobile-devices-rule-over-pcs-for-social-access/.

Maras, S. (2009). *Objectivity in Journalism*. Cambridge: Polity Press.

Mazzoleni, G. (1987). Media logic and party logic in campaign coverage: The Italian general election in 1983. *European Journal of Communication*, 2 (1), 81–103.

McManus, J. (2009). The Commercialization of News. In: K. Wahl-Lorgensen and T. Hanitzsch (eds.). *The Handbook of Journalism Studies*. New York; Routledge, 218–235.

McQuail, D. (2013). *Journalism and society*. London: Sage.

McQuail, D. (2010). *McQuail´s Mass Communication Theory*, Sixth edition, London: Sage.

Media use in the European Union. Report. Standard Eurobarometer 78 Autumn 2012. Retrieved July 15, 2014 from http://ec.europa.eu/public_opinion/archives/eb/eb78/eb78_media_en.pdf.

Meyers, C. (ed.) (2010). *Journalism Ethics: A Philosophical Approach*. Oxford: Oxford University Press.

Milewski, N., P. Barczyszyn and E. Lauk (2014). Three countries, one profession: the journalism cultures in Poland, Romania and Moldova. In: Głowacki, M. E. Lauk, A.Balčytienė (eds.). *Journalism that Matters. Views from Central and Eastern Europe*. Frankfurt am Main: Peter Lang Edition, 97–114.

Miller, D., and W. Dinan (2009). Journalism, Public Relation and Spin. In: K. Wahl-Lorgensen and T. Hanitzsch (eds.). *The Handbook of Journalism Studies*. New York and London; Routledge, 250–264.

Mirsky, D. (1999). *The History of the Russian Literature from Its Beginning to 1900*. Evanston, IL: Northwestern University Press.

Mitchelstein, E., and P.J. Boczkowski (2009). A review of recent research on online news production. *Journalism*, Vol. 10 (5), 562–586.

Nikunen, K. (2013). Losing my profession: Age, experience and expertise in the changing newsrooms. *Journalism online*. Retrieved April 25, 2014 from http://www.uta.fi/cmt/tutkimus/comet/tutkimus/paattyneet_projektit/lama/Losing%20my%20profession.full.pdf.

NK.pl straciła najwięcej młodych, wykształconych i wielkomiejskich użytkowników (analiza) [NK.pl lost the youngest, educated and metropolitan users (analysis)]. Retrieved June 6, 2014 from http://www.wirtualnemedia.pl/artykul/

nk-pl-stracila-najwiecej-mlodych-wyksztalconych-i-wielkomiejskich-uzytkownikow-analiza.

Nord, L. (2008). Comparing Nordic media systems: North between West and East? Central European Journal of Communication. Vol. 1, 95–110.

Nord, L. and M. Grusell (2012) *Inte för smalt, inte för brett: spelet om framtidens public service* [Not to broad, not to narrow: the game about future for public service]. Göteborg: Stiftelsen Etermedierna i Sverige in cooperation with NORDICOM.

Nordenstreng, K., Topuz, H. (eds.). (1989). *Journalist: Status, Rights and Responsibilities.* Prague: International Organization of Journalists.

Novikova, A.A. and I. Kiriya (2013). Depressivnoe mediapotreblenie (issledovanie televizionnyh predpochtenij sel'skih zhitelej) [Depressive media consumption. Research of TV preferences of rural inhabitants]. *Vestnik Moskovskogo universiteta. Serija 10: Zhurnalistika* [Vestnik MSU, Series 10 Journalism], 6, 154–167.

Nygren, G. (2014). *Multiskilling in the newsroom – de-skilling or re-skilling of journalistic work?* Journal of Media Innovation Vol. 1 (2), 75–96.

Nygren, G. (2013a). Changing journalistic culture. In: M. Anikina, B. Dobek-Ostrowska and G. Nygren (eds.). *Journalists in Three Media Systems. Polish, Russian and Swedish journalists about values and ideals, daily practice and the future.* Moscow: Moscow State University, 5–11.

Nygren G. (2013b). Professionalization of journalism. In: M. Anikina, B. Dobek-Ostrowska and G. Nygren (eds.). *Journalists in Three Media Systems. Polish, Russian and Swedish journalists about values and ideals, daily practice and the future.* Moscow: Moscow State University, 12–29.

Nygren, G. (2012a). The research project Journalism in change. In: G. Nygren, G.(ed.). *Journalism in Russia, Poland and Sweden – traditions, cultures and research.* Stockholm: Södertörns University, 5–12.

Nygren, G. (2012b). Autonomy – a crucial element of professionalization. In: G. Nygren, G.(ed.). *Journalism in Russia, Poland and Sweden – traditions, cultures and research.* Stockholm: Södertörns University, 73–95.

Nygren, G. (ed.) (2012c). *Journalism in Russia, Poland and Sweden – traditions, cultures and research.* Stockholm; Södertörns University.

Nygren, G. (2011). Passing Through Journalism. In B. Franklin and D. Mensing (eds.) *Journalism Education, Training and Employment.* London; Routledge, 207–221.

Nygren, G. (2008a). *Nyhetsfabriken – journalistiska yrkesroller i en förändrad medievärld.* [The News Factory – journalistic professional roles in a changed media world]. Lund: Studentlitteratur.

Nygren, G. (2008b). *Yrke på glid: Om journalistrollens deprofessionalisering* [Occupation on the skids: the de-professionalization of journalism]. Stockholm: Institutet för mediestudier.

Nygren, G. and J. Appelberg (2013). Swedish journalists – a profession in decline? In: M. Anikina, B. Dobek-Ostrowska and G. Nygren (eds.). *Journalists in Three Media Systems. Polish, Russian and Swedish journalists about values and ideals, daily practice and the future.* Moscow: Moscow State University, 115–164.

Nygren, G. and E. Appelgren (2014) Data Journalism in Sweden. *Digital Journalism* 2:3, 394–405.

Nygren G., B. Dobek-Ostrowska and M. Anikina (2013). Towards new questions. In: M. Anikina, B. Dobek-Ostrowska and G. Nygren (eds.) (2013). *Journalists in Three Media Systems. Polish, Russian and Swedish journalists about values and ideals, daily practice and the future.* Moscow: Moscow State University, 165–172.

Nygren, G. and E. Degtereva (2012). Russian and Swedish journalists. Professional roles, ideals and daily reality. *Journalism Practice.* Vol. 6, 5–6. DOI:10.1080/17512786. 2012.667277.

Nygren, G., E. Degtereva and M. Pavlikova (2010). Tomorrow's journalists. Trends in the development of the journalistic profession as seen by Swedish and Russian students. *Nordicom Review.* Vol. 31 (2), 113–133.

Nygren, G and K. Stigbrand (2013). The formation of a professional identity. Journalism students in different media systems. *Journalism Studies,* DOI:10.1080/146 1670X.2013. 834163.

Nygren, G. and M. och Zuiderveld (2011). *En himla många kanaler – flerkanalpublicering i svenska mediehus* [A lot of channels – cross platform publishing in Swedish media companies]. Göteborg: Nordicom.

Oates, S., G. McCormack (2010). The Media and Political Communication. In: S. White (ed.). *Developments in Russian Politics* 7. Durham: Duke University Press.

Ottosen, R. (2004) *I journalistikkens grenseland – journalistrollen mellon marked og idealer.* Kristiansand: IJ-forlaget.

Örnebring, H. (2010a) Technology and journalism-as-labour: Historical perspectives. *Journalism,* 11 (1), 57–74.

Örnebring, H. (2010b) Reassesing Journalism as a Profession. In S. Allan (ed.). *The Routledge Companion to News and Journalism*. London and New York: Routledge, 568–577.

Örnebring, H. (2009). *The Two Professionalisms of Journalism*. Oxford: Reuters Institute for the Study of Journalism. Retrieved September 25, 2011 from http://reutersinstitute.politics.ox.ac.uk/fileadmin/documents/Publications/The_Two_Professionalisms_of_Journalism_Working_Paper.pdf.

O'Sullivan, J. A. och Heinonen (2008) Old values, new media. *Journalism Practice* 2 (3), 357–371.

Pasti, S. (2014). Russian journalism as a social lift: comparing journalistic attitudes in the period 1992–2008. In: M. Głowacki, E. Lauk, and A. Balčytienė (eds.). *Journalism that Matters. Views from Central and Eastern Europe*. Frankfurt am Main: Peter Lang Edition, 153–174.

Pasti, S. (2013). *Alternative Journalism in the Russian Media System*. Paper presented on the Workshop "Russian and East European media: Perspectives on Consumption and Social Change" (The Nordic network on Choices, Resources and Encounters in Russia and other European post-socialist states – CERES), organized by Aleksanteri Institute, Helsinki, Finland, May 16–17, 2013. http://www.helsinki.fi/aleksanteri/ceres/workshop_media.html.

Pasti, S. (2012). Sovremennye rossijskie zhurnalisty: otnoshenie k professii [Contemporary Russian journalists: relation toward profession]. *Zhurnalistika*. Moscow: Moscow State University, 22–41.

Pasti, S. (2011). Sovremennye rossijskie zhurnalisty: otnoshenie k professii [Contemporary Russian Journalists and their attitudes to the profession]. *Vestnik Moscow State University, Zhurnalistika*, 4 July-August, 35–36.

Pasti, S. (2007). *The Changing Profession of a Journalist in Russia. Academic dissertation*. Tampere: University of Tampere, Department of Journalism and Mass Communications, Finland.

Pasti, S. (2005). Two Generations of Contemporary Russian Journalists, *European Journal of Communication*, 20(1): 89–115.

Pasti, S., M. Chernysh, and L. Svitich (2012). Russian Journalists and Their Profession. In: D. Weaver and L. Willnat (eds.). *The Global Journalist in the 21st Century*. New York and London: Routledge, 267–282.

Pasti, S., and K. Nordenstreng (2013) Paradoxes of journalistic profession: Case of Russia in the context of the BRICS countries. *World of Media 2012 Yearbook of Russian Media and journalism Studies*, Moscow: Media Mir, 243–268.

Pasti, S. and J. Pietiläinen (2008). Journalists in the Russian Regions: How different Generations View their Professional Roles. In: S. White (ed.). *Media Culture and Society in Putin's Russia*. London: Palgrave Macmillan, 109–132.

Peters, C. and M. Broersma (2012).*Rethinking journalism: trust and participation in a transformed news landscape*, London: Routledge.

Peters, J., E.C. Tandoc, (2013). People who aren't really reporters at all, who have no professional qualifications: Defining a Journalist and Deciding Who May Claim The Privileges, *N.Y.U. J. LEGIS. & PUB. POL'Y QUORUM* (34), 34–63.

Peters, J. (2013). *Why Defining a Journalist Is Messy But Crucial?* Retrieved January 20, 2014 from http://www.pbs.org/mediashift/2013/10/why-defining-a-journalist-is-messy-but-crucial/.

Philips, A. (2012). Faster and shallower: homogenisation, cannibalisation and the death of reporting. In: P. Lee-Wright, A. Philips, T. Witschge (eds.) *Changing Journalism*. London and New York: Routledge, 81–98.

Podshibjakin, A. (2010). *Po zhivomu. 1999–2009: LiveJournal v Rossii*. [In living. 1999–2009: LiveJournal in Russia.] Moscow: Izdatelskaja Gruppa Attikus.

Press Freedom Index 2013 Retrieved April 12, 2014 from http://en.rsf.org/press-freedom-index-2013,1054.html.

Podshibiakin, A. (2010). *Po-zhivomu. 1999–2009. LiveJournal v Rossii* [In living. 1999–2009: LiveJournal in Russia]. Moscow: Attikus.

Prokhorov, E. (2007). *Vvedenie v teoriyu zhurnalistiki* [Introduction to theory of journalism]. Moscow: Aspekt Press.

Quandt, T. and Singer J. B. (2009). Convergence and Cross-Platform Content Production, in Hanitzsch/Wahl-Jorgensen (eds.). *The Handbook of Journalism Studies*. New York and London; Routledge, 130–145.

Radu. R, and D.A. Popa, (2014). Media Accountability and Journalism Education. In: S. Fengler, G. Mazzoleni, S. Russ-Mohl, C. Porlezza, T. Eberwein (eds.). *Journalists and Media Accountability: An International Study of News People in the Digital Age*. New York: Peter Lang, 249–263.

Reich, Z. and T. Hanitzsch (2013). Determinants of Journalists Professional Autonomy: Individual and National Level Factors Matters More Than Organizational Ones. *Mass Communication and Society*, 16, 133–156.

Reporters without Borders. For freedom of information. Retrieved February 12, 2014 from http://en.rsf.org/.

Roudakova, N. (2012) Comparing Processes: Media, Transitions and Social Change. In: D.C. Hallin, and P. Mancini. (eds.). *Comparing Media Systems Beyond the Western World.* Cambrige University Press, 246–277.

Roudakova, N. (2009). Journalism as "Prostitution": Understanding Russia's Reactions to Anna Politkovskaya's Murder. *Political Communication*, 26 (4), 412–429.

Russian Union of Journalists. Retrieved April 21, 2014 from http://www.ruj.ru.

Schudson, M. (2011). *The sociology of news. 2., [rev. and updated].* New York: W.W. Norton and Co.

Schudson, M. (2003). *The Sociology of News*, New York: WW Norton and Company.

Schudson, M. (2001). The objectivity norm in American journalism. *Journalism.* 2, 149–170.

Scott, J., and G. Marshall (2005). *Oxford dictionary of sociology.* Oxford University Press.

Shoemaker, P.J. and S.D. Reese (1996). *Mediating the message: Theories of influences on mass media content.* New York: Longman Publishers.

Smith, A., and J. Brenner (2012). *Twitter Use 2012. Report.* Pew Research Internet Project. Retrieved May 7, 2014 from http://www.pewinternet.org/2012/05/31/twitter-use-2012/.

Singer, J. B. (2010). Norms and the Network: Journalistic Ethics in a Shared Media Space. In: Ch. Meyers (ed.). *Journalism Ethics: A Philosophical Approach.* Oxford University Press, 117–130.

Singer, J. B. (2005). The Political J-blogger: 'Normalizing' a New Media Form to Fit Old Norms and Practices. *Journalism.* Vol. 6 (2), 173–198.

Singer, M. G. (1961). *Generalization in Ethics – An essay in the Logic of Ethics, with the Rudiments of a System of Moral Philosophy.* New York: Atheneum.

Singer, J. B.; A. Hermida, D. Domingo, A. Heinonen, S. Paulussen, T. Quandt, Z.Reich, M. Vujnovic (2011). *Participatory Journalism. Guarding Open Fatesg at Online Newspapers.* Chichester: Wiley-Blackwell.

Social Networking Popular Across Globe Arab Publics Most Likely to Express Political Views Online. Survey Report. December, 2012. Retrieved May 7, 2014 from http://www.pewglobal.org/2012/12/12/social-networking-popular-across-globe/.

Sosnowska. J. (2013). *To już koniec Blipa – serwis będzie działać do końca wakacji* [This is the end of Blip – the service will operate until the end of the summer holidays]. Retrie- ved June 6, 2014 from http://technologie.gazeta.pl/internet/1,104530,14200685,To_juz_koniec_Blipaserwis_bedzie_dzialac_do_konca.html.

State of the News Media 2014 Pew Institute. Retrieved March 22, 2014 from www.journalism.org/ packages/state-of-the-news-media-2014.

Stępińska, A., and M. Głowacki (2014). Professional Roles, Context Factors and Responsibility Across Generations of Polish Journalists. In: M. Głowacki, E. Lauk and A. Balčytienė (eds.). *Journalism that Matters: Views from Central and Eastern Europe*. Peter Lang, 79–96.

Stępinska, A., S. Ossowski, L. Pokrzycka, J. Nowak (2012). The Journalists and Journalism of Poland. In: D. Weaver and L. Willnat (eds.). *The Global Journalist in the 21st Century.* New York and London: Routledge, 255–267.

Stigbrand, K., and G. Nygren (2013). *Professional Identity in Changing Media Landscapes*. Stockholm: Södertörns University.

Stromback, J., L. Nord, and A. Shehata (2012). Swedish Journalists: Between Professionalization and Commercialization. In: D. Weaver and L. Willnat (eds.). *The Global Journalist in the 21st Century.* New York and London: Routledge, 309–319.

Strömbäck, J., and L. Nord (2006). Do Politicians Leads the Tango ?: A Study of the Relationship Between Swedish Journalists and their Political Sources in the Context of Election Campaigns. *European Journal of Communication*. Vol. 21, 147–164.

Strovski, D. (2012). Journalism in Russia. Why its future looks uncertain. In: E. Vartanova (ed.). *World of Media. Yearbook of Russian Media and Journalism Studies 2012*. Moscow: MediaMir, 297–312.

Sveningsson, M. (2013). Socialitet och socialt samspel på nätet. [Sociality and social interaction online.] In G. Nygren and I. Wadbring (eds.), *På väg mot medievärlden 2020: journalistik, teknik, marknad.* [On the way to media world 2020: journalism, technology, market.] Lund: Studentlitteratur, 249–268.

Szot, L. (2009). The Polish Journalism Education Landscape. In: G. Terzis (ed.). *European Journalism Education*. Bristol and Chicago: Intellect, 463–477.

Szynol, A. (2008). 15 years of Passauer Neue Presse and Orkla Media presence on Polish regional market. In: B. Dobek-Ostrowska and M. Głowacki (eds.). *Comparing Media Systems in Central Europe. Between Commercialization and Politicization*. Wroclaw; University of Wroclaw Press, 135–148.

TNS Gallup Media. Retrieved May 7, 2014 from http://www.tns-global.ru/services/media/media-audience/internet/information/?arrFilterpf%5BYEAR%5D=2012&set_filter=%D0%9F%D0%BE%D0%BA%D0%B0%D0%B7%D0%B0%D1%82%D1%8C &set_filter=Y.

Trakhtenberg, A. (2007). Transformation without change: Russian media as a political institution & cultural phenomenon. In: E. Vartanova (ed.). *Media and Change.* Moscow: MediaMir, 122–130.

Tuchman, G. (1983). *Making News, a Study in the Construction of Reality.* New York: The Free Press.

Tumber, H. and M. Prentoulis (2005). Journalism and the making of a profession. In H. de Burgh (ed.). *Making journalists.* London and New York: Routledge, 58–73.

Tunstall, J. (1971). *Journalists at work. Specialist correspondents: their news organizations, news sources, and competitor-colleagues.* Beverly Hills: Sage.

Twitter Now The Fastest Growing Social Platform in The World. GlobalWebIndex Blog. 2013-01-28. Retrieved May 7, 2014 from http://blog.globalwebindex.net/twitter-now-the-fastest-growing-social-platform-in-the-world/Re.

Unicos blog. World map of social networks. Retrieved May 7, 2014 from http://vincos.it/world-map-of-social-networks/.

Vartanova, E. (2012). The Russian media Model in the Context of Post-Soviet Dynamics, In: D. C. Hallin, and P. Mancini (2012). *Comparing Media Systems Beyond the Western World.* Cambridge: Cambridge University Press, 119–142.

Vartanova, E. (2007). Russian Media Economy: Euroasian model. In: E.Vartanova (ed.). *Media and Change.* Moscow: MediaMir, 103–121.

Vartanova, E. and Azhgihina, N. (2011). *Dialogi o zhurnalistike* [Dialogs about journalism] Moscow: Fakultet zhurnalistiki MGU imeni M. V. Lomonosova.

Vartanova, E., M. Lukina, L. Svitich and A. Shiryaeva (2010). Between Tradition and Innovation: Journalism Education in Russia. In: B. Josephi (ed.). *Journalism Education in Countries with Limited Media Freedom.* New York: Peter Lang Publishing, 199–216.

Volek, J. (2010). Czech Journalists after the Collapse of the Old Media System: Looking for a New Professional Self-Image. In: B. Dobek-Ostrowska, M. Głowacki, K. Jakubowicz, M. Sükösd (eds.). *Comparative Media Systems. European and Global Perspective.* Budapest – New York: CEU Press, 171–194.

Von Krogh, T. (2008). *Mediernas ansvarighet: om media accountability på svenska* [The responsibility of media; on media accountability in Swedish]. Stockholm: SIMO- förlag.

Voronova, L. (2014) *Gendering in Political Journalism. A Comparative Study of Russia and Sweden*. Örebro: Örebro University.

Wadbring, I. (2012*). Journalistikens kommersialisering – mera myt än sanning?: innehållets förändring i svensk dagspress 1960–2010* [Commercialization of journalism – more myth than truth?]. Stockholm: Stiftelsen Institutet för mediestudier (Sim(o)).

Wahlund, R., C. Rademaker, P. Nilsson, and M. Svahn (2013). Mediernas roll I marknadskommunikationen [The role of media in market communication]. In: G. Nygren and I Wadbring (eds.). *På väg mot medievärlden 2020* [On the road to Media world 2020] Lund: Studentlitteratur.

Waisbord, S. (2013). *Reinventing professionalism: Journalism and News in Global Perspective.* Cambridge, UK: Polity Press.

Waisbord, S. (2010). Rethinking "Development" Journalism. In: S. Allan (ed.). *The Routledge Companion to News and Journalism.* London and New York: Routledge, 148–158.

Ward, S. J. A. (ed.). (2013). *Global Media Ethics: Problems and Perspectives.* New York: Wiley-Blackwell.

Ward, S.J.A. (2008). Global journalism ethics: Widening the conceptual base. *Global Media Journal – Canadian Edition/* 1(1), 137–149.

Weaver, D. and L. Willnat (2014). *The American Journalist in the Digital Age. Key Findings.* Bloomington: School of Journalism, Indiana University.

Weaver, D. and L. Willnat (eds.) (2012). *Global Journalist in the 21st Century.* New York: Routledge.

Weaver, D. and W. Wu, (1998). *The Global Journalist. News People Around the World.* New Jersey: Hampton Press.

Weaver, D., Beam, A. Randal, Brownlee, J Bonnie; Paul S Voakes; Wilhoit, G Cleveland (2007). *The American Journalist in the 21st Century.* New Jersey: Lawrence Erlbaum.

Weekly survey "FOMnibus" March, 22–23, 2014 (2014). Retrieved: April 7, 2014 from www.fom.ru/posts/download/11427.

White, S. (2011). *Understanding Russian Politics.* Cambridge: Cambridge University Press.

Who is a journalist? Retrieved January 12, 2014 from http://www.udayton.edu/news/articles/2013/10/who_is_a_journalist.php.

Wiik, J. (2012). Yrkesideal Occupational ideals]. In K. Asp, (ed.). *Svenska journalister: 1989–2011* [Swedish Journalists 1989–2011]. Göteborg: Institutionen för journalistik, medier och kommunikation (JMG), Göteborgs universitet.

Wiik, J. (2010). *Journalism in transition: the professional identity of Swedish journalists*. Göteborg: Department of Journalism, Media and Communication, University of Gothenburg (JMG). Retrieved February 20, 2014 from http://hdl.handle.net/2077/21853.

Witschge, T., and G. Nygren, G. (2009). Journalism: A Profession Under Pressure? *Journal of Media Business Studies*, 6 (1), 37–59.

Wilkins, L. and C. G. Christians (eds.) (2009). *The handbook of mass media ethics*. London: Routledge.

World Internet Project. International Report. Fifth Edition (2013) University of Southern California, Center for the Digital Future USC Annenberg School for Communication and Journalism. Retrieved June 6, 2014 from www.worldinternetproject.net.

World Map of Social Networks. Unicos blog. Retrieved May 7, 2014 from http://vincos.it/world-map-of-social-networks/.

World Values Survey Wave 6: 2010–2014, Retrieved April 17, 2014 from http://www.worldvaluessurvey.org/WVSOnline.jsp.

Wu, W., D. Weaver, J. Owen, and J. Johnstone. (1996). Professional rules of Russian and IS journalists: a comparative study. *Journalism and Mass Communication Quarterly* 73 (3), 534–548.

Zelizer, B. (ed.) (2009). *The Changing Faces of Journalism: Tabloidization, Technology and Truthiness*. London and New York: Routledge.

Zelizer, B. (2005). The culture of journalism. In: J. Curran and M. Gurevitch (eds.). *Mass media and society*. London: Hodder Arnold.

Zelizer, B. (2004). *Taking Journalism Seriously. News and the Academy*. Thousand Oaks: Sage.

Zhirkov, G. V. (2001). *Istorija cenzury v Rossii XIX–XX vv.* [History of censorship in Russia in XIX–XX]. Moscow: Aspekt Press.

List of tables

Chapter 6

Chapter 7

Chapter 8

Chapter 9

Chapter 10

List of figures

Chapter 8

Appendix

Appendix 1: The Questionnaire of the "Journalism in change" Survey, 2012

Stockholm/Moscow/Wrocław February 2012

Dear Journalist,

New technology, commercial pressure and an evolving network society changes the conditions for journalism – but is it changing in the same way in different media systems? How are journalistic practices and journalistic values influenced by media development in different countries?

The purpose of the project "Journalism in change" is to study professional journalistic cultures in three countries: Sweden, Poland and Russia. These countries have a different historical and political legacy, the relations between media and political/economic power are quite different and there are big differences in media use. Our question is how media development is influenced by the different conditions, and how professional journalistic cultures are changing in these three countries.

In the project, researchers from Södertörn University in Sweden, Wrocław University in Poland and Moscow State University in Russia are working closely together to find some answers. One of the most important steps is this survey to 500 journalists in each country, a sample chosen to be representative for the journalistic population in the country.

We hope that you want to participate in the project by giving your answers to the questions in the survey. If you find any question difficult to answer, just leave it and answer the rest. Every answer is important to give us a reliable picture of how journalists perceive their daily work and professional values.

If you have any questions, don't hesitate to ask. Send us an e-mail and we will get back to you as soon as possible.

When you have filled in the survey, put it in the envelope and give it to the person distributing the surveys. All answers are anonymous, it will not be possible for neither the researchers nor anyone else to see who has given individual answers.

If you want to take part in the forthcoming interviews or get the results from the survey, please send us an e-mail. The first results are planned to be published during 2013.

Gunnar Nygren,
Södertorn University
Email: gunnar.nygren@sh.se

Boguslawa Dobek-Ostrowska
Wrocław University
Email: dobek@uni.wroc.pl

Maria Anikina
Moscow State University
E-mail: maria-anikina@yandex.ru

A: General questions

Q1. *Your gender*

- Female
- Male

Q2. *What year were you born?:*

Q3. *In what kind of media company are you working today?*
(If you work for different kinds of companies, please give the main option)

- National newspaper/big city newspaper
- Regional newspaper/local newspaper
- Free newspaper
- Popular/weekly magazine
- Specialized magazine
- Other newspaper/magazine
- Production company for print
- Online publishing company
- TV-channel
  - public service/state
  - commercial
- Radio Station
  - public service/state
  - commercial
- Production company for radio/TV/film
- News agency
- Other, please specify

Q4. *How many hours do you work as a journalist in an average week?*
(One option)

- I work full-time as a journalist, approximately … hours/week
- I work part time as a journalist approximately … hours/week, and part time with other kind of work.

Q5. *What kind of main employment do you have today*)?
(One option)

- Regular employment in a media company
- Temporary employment (filling in, project etc.)
- Temporary job agency (like Manpower)
- Freelance/your own company (go to question 8)
- Other; please specify…

Q6. *How many full-time news and editorial people are employed in your organization (your place of work)*
… number of full-time journalists.

Q7. *During the last five years, has the size of your editorial staff grown or shrunk?*

- Shrunk substantially
- Shrunk a little
- The same
- Grown a little
- Grown substantially
- Don't know
  Your comments…

Q8. *Are you a member of any professional organizations for journalists?*
(You can choose many alternatives)

- Union of journalists
  Name…
- Other kind of professional organizations
  Name…
  Name…
  Name…

Q9. *For how many years have you worked as a journalist?*

Q10. *Where would you like to be working in five years from now?* (One alternative)

- In the same media company at present
- In some other media company

What kind of media?
- In some other field than journalism.

What other field?
- I will be retired
- Don't know

Q11. *How important are a number of things in choosing jobs for you?*

Please rate on a scale from 1 (Not important) to 5 (Very important), I do not know

- Payment/salary
- Fringe benefits
- The editorial policy of the organization
- Job security
- The chance to develop a specialization
- The atmosphere in the workplace
- To serve the public interest and to help people
- The amount of freedom in work
- The chance to make a better career in journalism
- The chance to influence society
- Other; what:
  Your comment, if you like to add:

B. Your daily work

Q12. *What are your main functions in your media organization*? (Maximum three alternatives)

- Reporter/correspondent
- Copy editor/editor/layout
- Photographer/cameraman
- Researcher
- Analyst/commentator
- Editorial writer/columnist
- Producer (radio/TV)

- Anchor person (radio/TV)
- Online publishing
- Editor/managing other journalists
- To work with advertisements
- Economic matters in the company
- Other, please specify…

Q13. *Do you usually cover one particular subject area, or do you cover different things?*

- I cover a particular area or subject (go to question 14)
- I cover different topics and subjects (go to question 15)

Q14. *If you cover a specific topic or subject, in what area?*
(Choose only one alternative)

- Crime and law
- Culture
- Economy/business
- Entertainment/glamour
- Family/human stories
- Foreign news
- Lifestyle
- Politics
- Science
- Society and social issues
- Sports
- Other, please specify…

Q15. *For what platforms are you producing/editing content in an average week?* (You can choose as many options as you need)

- Print
- Television
- Radio
- Online
- Mobile devices

Q16. *If you have a good idea for a subject which you think is important and should be followed up, how often are you able to get the subject covered?* (Only one option)

- Almost always
- More often than not

- Only occasionally
- I don't make such proposals
- Don't know

Q17. *Some questions about your daily work*:
Please rate on the scale 1 (No freedom at all), 2 (Only some freedom), 3 (A great deal of freedom), – 4 (Almost complete freedom), I do not know.

- How much freedom do you have in selecting the stories you work on?
- How much freedom do you usually have in deciding which aspects of a story should be emphasized?
- How much freedom do you have to manage your own time in your work?

Q18. *Different factors influence the selection of subjects in the daily work of your media organization. How could you evaluate the significance of the following factors in the work on your news organization?*
Give your answer on a scale 1 (insignificant) -5 (very significant)

- Professional interest of journalist
- The editorial policy of the media company
- The political connection (affiliation) of the media company
- Strong political actors outside the media company
- Audience interest
- Needs and interests of society
- Those who advertise in the media
- Economic power outside the media company (big companies etc.)
- Other (please specify)
  Your comment if you like to add…

Q19. *Various methods are used to influence journalists and the output of media companies. How often do you deal with the following methods of influence at your work place?*
Give your answers on a scale 1 (Never) to 5 (Very often), I do not know

- Press material delivery, e.g., text and pictures/film
- Press conferences
- Personal information for journalists selected by the sources
- Information from agents on behalf of a hidden source
- Gifts to the journalists (also travels etc.)
- Pressure and threats against an individual journalist

- Pressure and threats against the media company
- Other (please specify)
  Your comment if you like to add…

Q20. *How often does your editor/chief change anything in what you produce (text, pictures, plots)*
Give your answers on a scale 1 (Almost always), 2 (More often than not), 3 (Only occasionally), 4 (Almost never), I do not know
Q21. *A journalist can face various obstacles before the final publishing/broadcast. How often do journalists at your media company face the following obstacles?*
Range your answers on a scale 1 (Never) -5 (Very often), I do not know

- Limits in the editorial policy of the media company
- Political interests of the owners of the media company?
- Economic interests of the media company
- Pressure from political actors outside the media company?
- Pressure from advertisers
- Pressure from the sources of information
- Other (please specify)
  Your comment if you like to add…

Q22. *During the last 5–10 years daily journalistic work has changed in many ways. How do you evaluate these changes in different areas, according to your own experience?*
Range your answers on a scale 1 (Much less) -5 (Much more), I do not know, I don't have such extensive experience

- Work outside the newsroom
- Research and verification
- Writing/producing
- Editing/layout
- Contacts with the audience
- Cooperation in the newsroom
- The amount of production
- Evaluation and discussion
- Photography
- Other, please specify…

Q23. *The need to verify and the need to publish news as fast as possible are two basic journalistic values that can be in conflict with each other. Do you agree or not agree with the following statements?*
Range your answers on a scale 1 (Disagree) -5 (Fully agree), I do not know

- In my news organization, news is equally verified in all channels (print, online, radio, TV).
- Incorrect facts are often published due to too little control.
- It is better to publish unverified information than to be slower than our competitors.
- The audience has lower demands on verification in online news than in our main channel (newspaper, TV or radio)
- Online journalism is more of a process, and the verification can be done during the process and not before publishing.

Q24. *"Multiskilling" or "multi-reporter" are some words often used to describe future journalists. It can mean both a journalist working in different channels and a journalist working with different parts of the process (photo-writing-editing). What is your opinion on the following statements?*
Range your answers on a scale 1 (Disagree) -5 (Fully agree), I do not know

- I feel like a multi-skilled reporter
- Journalists in my media organization are expected to be multiskilled
- Multiskilling means more space for creativity
- Multiskilling will decrease the quality of journalism
- Multiskilling gives more power to the individual journalist
- In the future more journalists will be multiskilled
  Your comments if you like to add:

Q25. *Digital technology has changed daily journalistic work over the last 10–15 years. What is your opinion on the following statements on the influence of technological changes?*
Range your answers on a scale 1 (Disagree) -5 (Fully agree), I do not know

- New technology …
- makes the work process faster
- makes the work more creative
- increases time pressure in daily work

- has a positive potential not used in daily work
- creates a lot of technological problems
- puts a focus more on form and production, than on journalistic content

Q26. *If you look back 5–10 years, how would you evaluate the changes in influence on media content in your news organization from the following actors?*

Range your answers on a scale 1 (Decreased) -5 (Increased), I do not know

I don't have such extensive experience

- Journalists
- Other groups in the media company
- Media owners
- Politicians
- Audiences
- Interest groups
- Advertisers
- Private business
- Other

Q27. *Here are some statements about your relation to your profession. Do you agree or disagree with the statements?*

Range your answers on a scale 1 (Disagree) -5 (Fully agree), I do not know

- I am proud to call myself a journalist
- My profession is the mission in my life
- The main reason for working as a journalist is to get a salary
- My work means more for me than family and leisure time
- Working as a journalist is a step in my career to other areas

Q28. *Here are some statements on journalists, ethics and politics. Do you agree or not?*

Range your answers on a scale 1 (Disagree) -5 (Fully agree), I do not know

- A journalist cannot himself be active in politics at the same time as he is working as a journalist.
- It is no problem for a journalist to work with PR and information
- As a journalist, it is OK to receive gifts from sources as long as it does not influence my professional work.

- It is very important to have common ethical rules among journalists in the media company.
- As a journalist one has to accept political influence in the media company
  Your comments if you like to add:

C. Journalists, journalism and society

Q29. *How important are the following professional duties for a journalist in your country?* (Give your opinion on a scale 1 (Not important) -5 (Very important), I do not know

Journalists should…

- criticize injustice
- tell the truth regardless of the consequences
- ensure that the media business does well
- be a neutral reporter
- stand free of special interests
- bring forward various opinions
- simplify and explain
- influence public opinions
- provide entertainment
- investigate government claims
- provide information objectively
- control public authorities
- represent diverse social groups
- mobilize people to act
- educate the public
- stimulate new ideas

Q30. *Do you think that journalists in your country fulfil their responsibilities towards society?*

(Give your answer on a scale 1 (No) -5 (Yes), I do not know
Your comment if you'd like to add:.

Q31. *Do you think the level of press freedom in your country has increased or decreased over the last 5–10 years?*

Give your answer on a scale: 1 (Decreased), 2 (Stayed the same) 3 (Increased), I don't know.
Your comment if you'd like to add:

Q32. *Do you think the quality of journalism in your country has increased or decreased over the last 5–10 years?*
Give your answer on a scale: 1 (Decreased), 2 (Stayed the same) 3 (Increased), I do not know.

Your comment if you'd like to add:

Q33. *The independence of journalism can be influenced by different factors. How do you evaluate the influence of the following factors in your country?*
(Give your opinion on a scale 1 (No danger) -5 (A great danger), I do not know

- The owners demand for profits
- Increasing work tempo
- Foreign ownership
- Threats against individual journalists
- State influence (laws and ownership)
- Advertisers' influence on content
- Political influence in media companies
- Weak professional ethics
- Source activity (PR/lobbyists)

Q34. *What kind of future awaits journalism as a profession in XX in your opinion?* (Only one alternative)

- The profession will flourish
- It will remain the same as today
- The profession will vanish step by step
- The profession will transform to another activity
- I don't know

Your comment if you like:

D. Journalism and interactivity

Q35. *How often do you communicate with readers/viewers/listeners?*
(Give your opinion on a scale 1 (Never), 2 (Sometimes), 3 (Every week), 4 (Every day), -5 (Many times a day), I do not know

- By telephone
- By e-mail
- With comments on the media website/blog
- With social media
- Directly by meeting people
- Other:…

Q36. *Interactivity is often discussed in connection to media development. Do you agree or disagree with the following statements?*

(Give your answers on a scale 1 (Disagree) -5(Fully Agree), I do not know

- Increased interactivity influences the selection of news
- It is important to listen to the opinions of the audience
- Increased interactivity can improve the quality of journalism
- Contacts with the audience detracts too much from work time
- A journalist should not listen too much to opinions among the audience

Your comments:

Q37. *Social media and blogs is a growing part of the media landscape. Do you use any type of social media in relation to your work (blogs, Facebook, Twitter or other communities)?*

Yes (go to question 38), No (go to question 40)

Q38. *How often do you use different types of social media?*

Give your opinion on a scale 1 (Never), 2 (Sometimes), 3 (Every week), 4 (Every day), -5 (Many times a day), I do not know

- Read blogs (not connected to media companies)
- Write in my own personal blog
- Write in a media blog on the website of a media company
- Use Facebook
- Use Twitter
- Use other kinds of communities

Your comments if you like:

Q39. *For what purposes do you use different kinds of social media? Facebook Twitter Blogs*

(You can choose as many options as you want)

- To get ideas
- In research/investigation (to find information)
- To have a dialogue with the audience
- To keep in contact with colleagues
- Because it is expected from the editors that I should
- To publish content besides my regular work
- To get more readers/viewers, self-promotion

- To strengthen the trademark of my media company
- For professional discussions
- For making money by advertisements or PR
- To discuss social and political questions in general
- Other purposes, please specify…

E. Some personal information

Q40. *What is the highest level of your education?*

- Primary school
- Secondary school
- University or other higher education

Q41. *Do you have a special education in journalism?*

- Yes, at an academic level
- Yes, at a non-academic level
- Yes, professional training in the media company
- No

Q42. *What is the highest education of your parents?*

My father

- Primary school
- Secondary school
- University/higher education

My mother:

- Primary school
- Secondary school
- University/higher education

Q43. *Do any of your relatives work (or have worked) as journalists?*

- Wife/husband
- Parents
- Children
- Other close relatives
- No

Q44. *Is it possible to manage on your income from journalism?*
(One alternative)

- Yes, I can do well and provide extra things for me
- Yes, I can provide all necessary things
- Yes, I can survive
- No, I also have to earn money from other jobs
- No, journalism is only a minor part of my income

Q45. *How interested are you in politics and political questions in society?*
(Give your answer on a scale 1 Not interested) – 5 (Very interested), I do not know

Q46. *Is it an alternative for you to be active in politics? (voluntary or as employed)*

- No
- Perhaps,
- Yes
- Don't know

Q47. *Where would you place yourself on a political scale from left to right?*

- Left
- A bit left
- Center
- A bit right
- Right
- Don't know
- I choose not to answer

If you have any other comments, please give them here:

Appendix 2: Interview

Presentation to the interviewee:

We are conducting a survey of 20 journalists by country, in Sweden, Poland and Russia. The survey aims to obtain a picture of professional everyday life in the three countries and also to enable a comparison between the countries.

I. Professional history – experience, training, why journalist

1. Why did you become a journalist?
2. What is your training and experience?
3. What is it that makes young people want to be a journalist today? Is there any difference from when you started?

II. Professional practice – daily work, multiskilling position

4. Please describe your last working day (tasks, how you interact with your work management, time given for different tasks, collaboration with colleagues)
5. What are the key skills of a journalist in your company / alternative for you as a freelancer?
6. What are the advantages and disadvantages of journalists being multi-skilled, that is, working with different media forms and with all parts of a production?
7. Which option do you have to work with a subject / material during a longer time span?
   (Follow up: the journalist works more with research today or less, does she/he have more/fewer resources available? Which are the main resources?)

III. Professional autonomy – pressure from inside and outside, limits for what is possible

8. What space do you have for your own initiatives and ideas? Please give examples.
9. How much of the source material that you use is prepared by others for publication (for example press releases)? What other methods are sources and actors outside your workplace using to influence your work?
10. Can you give examples of how your daily work and work in the newsroom is influenced by other units of the company:

a) in consideration of the advertising department?
b) in consideration to the financial or marketing departments?
c) in consideration of owners and their political orientation?

11. Is it possible for you to have any influence over your employment and your salary?
12. Are you a member of a trade union or other professional organization?

IV. Relationship to the audience

13. In what ways are you in contact with the audience? (contacts by phone and e-mail, comment columns on the Web, through social media and so on)
14. Do you often use social media (for example twitter, blogs, Facebook) in your work? Please give examples of how these media channels have been useful for you

V. Future – for individuals, for the profession

15. How do you think younger (alt. older) journalists look upon your generation of journalists?
16. What is your opinion about the similarities and differences between the generations?

If you look into the future; what are you doing in five years' time? In ten years? What will journalism in general be like in five years? In twenty years? How would you explain the changes?

Notes on contributors

Editors

Gunnar Nygren, Ph.D., is professor in journalism at Södertörn University, Sweden. He worked for a long time as a journalist. His research has been devoted to local journalism in Sweden and recent years mostly to professional roles of journalists and changing journalistic practices. He has participated in many national and international research projects. He was the leader for the transnational research project "Journalism in change. Journalistic culture in Poland, Russia and Sweden" (2011–2014) involving also researchers from Poland and Russia. He has published several articles on comparative studies on journalists and journalism students in Sweden, Poland, Russia and other Central and East European countries. E-mail: gunnar.nygren@sh.se

Boguslawa Dobek-Ostrowska, Ph.D., Professor at the University of Wrocław and Chair of Department of Communication and Journalism. She is a founder and past president (2007–2013) of the Polish Communication Association. She is a past president of the International Federation of Communication Associations (2010–2012). She has published several books and articles. She is interested in political communication and media systems. She is a founder and editor of the *Central European Journal of Communication*. She is a member of the editorial board of *Communication Theory*. She is an editor of the series Studies in Communication and Politics (Peter Lang Edition) and Politics and Communication (University of Wroclaw Press). E-mail: dobek@uni.wroc.pl

Contributors

Maria Anikina, Ph.D., associate professor at Lomonosov Moscow State University, Russia. She is the chair of the young scholars' council, a member of the academic council of the Journalism Faculty (MSU). She is an editorial board member of the "Kosciół i Media" series (Warsaw, Poland), chief editor of the electronic scientific journal "Mediascope". She participated in the research project "Media in Changing Russia" (2006–2008), and participated in the "The worlds of journalism" and "*Media systems in flux: The challenge of the BRICS countries*" projects. She is a member of the ICA, IAMCR, NAMMI. She has been awarded by Nevskaya premia (2013). Her research interests are media sociology, media and journalism studies. E-mail: maria-anikina@yandex.ru

Michał Głowacki, Ph.D., is an assistant professor at the University of Warsaw, Poland. In 2010, he was a Programme Advisor to the ad hoc Advisory Group on Public Service Media Governance (MC-S-PG) at the Council of Europe. He participated in the research project Media Accountability and Transparency in Europe (MediaAcT) funded by the 7th Framework Programme of the European Union (2010–2013). He has been awarded scholarships by the Danish Agency for Science, Technology and Innovation, the Swedish Institute, the Research Council of Norway, and the Polish-American Fulbright Commission. His research interests are media policy, accountability, and governance. e-mail: michal.glowacki@id.uw.edu.pl

Jöran Hök, Ph.D., senior lecturer at Södertörn University, Stockholm. He has earlier been working as a researcher and lecturer also at Stockholm and Gothenburg Universities in Journalism studies, in Mass communication research and also in Development studies and International relations. For ten years he was editor of the leading Swedish development journal „"OmVärlden" (in-the-World). He has also been an active journalist for more than 20 years in several Swedish daily newspapers. E-mail: joran.hok@sh.se

Elena Johansson, Ph.D., is a researcher at the School of Social Sciences, Södertörn University, Sweden. She was a journalist for national TV channels in Moscow, 1999–2006 and guest researcher at the Department of Journalism, Media and Communications, Stockholm University, Sweden, 2006–2009. Her research interests are technological changes in media (media convergence), social media, professional journalistic cultures, TV broadcasting. E-mail: elena.johansson@sh.se

Index

Studies in Communication and Politics

Edited by Bogusława Dobek-Ostrowska and Michał Głowacki

Vol. 1 Bogusława Dobek-Ostrowska / Jan Garlicki (eds.): Political Communication in the Era of New Technologies. 2013.

Vol. 2 Michał Głowacki / Epp Lauk / Auksė Balčytienė (eds.): Journalism that Matters. Views from Central and Eastern Europe. 2014.

Vol. 3 Gunnar Nygren / Bogusława Dobek-Ostrowska (eds.): Journalism in Change. Journalistic Culture in Poland, Russia and Sweden. 2015.

www.peterlang.com

www.ingramcontent.com/pod-product-compliance
Ingram Content Group UK Ltd.
Pitfield, Milton Keynes, MK11 3LW, UK
UKHW041825100926
8556IPUK00002B/26

9 783631 64921